Functional safety in Modern Mobility: ISO 26262 and Beyond

Dr. P. Arjunraj

ISBN 979-8-89475-337-9

The author, Dr. P. Arjunraj, extends his dedication of this book to every citizen of India in loving memory of the esteemed Dr. A. P. J. Abdul Kalam, Former President of India.

Contents

4. ISO26262 Part 3 – Concept Phase

5. Part 4 – Technical Safety Concept

SEooC – System Element out of Context

About the book

This book is an indispensable guide that navigates us through the marvelous landscape of automotive functional safety. In an era dominated by advanced electronic systems and functionalities, the book places Safety confirmed by the embedded systems at the forefront, offering a comprehensive understanding of this functional safety standard.

The book expertly blends theoretical concepts with practical applications, making it accessible to both novices and experienced professionals. It goes beyond the confines of ISO 26262 standard, delving into emerging trends and future challenges, providing a forward-looking perspective on the evolving automotive safety systems.

This book is well-structured for a diverse audience and is a valuable resource for academia, practitioners, and researchers. Its lucid style ensures that even complex topics are easily comprehensible, fostering a deeper understanding of automotive functional safety.

As the automotive industry hurtles towards a future marked by autonomy and electrification, this book serves as a compass, guiding readers towards safer and technologically advanced transportation solutions. It is a collaborative effort that mirrors the collaborative spirit necessary for success in this dynamic field.

Embark on an enlightening journey on automotive functional safety. Whether you are seeking to deepen your understanding of functional safety or staying ahead of future trends, this book is a beacon, illuminating the path towards a safer and more innovative automotive future.

This book covers the following 11 chapters

1. Introduction to Functional safety and standards

2. ISO26262 Part 1 Vocabulary

3. ISO26262 Part 2 Safety management

4. ISO26262 Part 3 Concept Phase

5. ISO26262 Part 4 Technical Safety Concept and SEooC

6. ISO26262 Part 9 ASIL decomposition

7. ISO26262 Part 4 Hardware Software Interface

8. ISO26262 Part12 Safety for Motorcycles

9. ISO 21448 Safety Of The Intended Functionality

10. Introduction to Automotive Cybersecurity

11. Functional Safety of Off-road vehicles

About the Author

Dr. P. Arjunraj is a Value engineering specialist and expert, best known for his Value engineering, Value analysis and Value Management expertise. Dr. P. Arjunraj is the author of the books "Concepts of Value Engineering" and "Practical Approach for Value Engineering using Tools and Techniques". He is a Certified Value Specialist (CVS) which is the highest level of certification in the field of Value Engineering certified by SAVE International, USA and Professional in Value Management (PVM) certified by European Value Management certification and training board. He is B.E Mechanical Engineering and Specialization Master of Engineering in Automobile from Anna University Chennai. He holds an M.B.A degree from Madras University specialized in Operations Management and completed his Ph.D. in B S Abdur Rahman University specialized in the field of Automobile Engineering.

He had published 10 Research papers in international journals, Presented 12 Research papers in international conferences and 10 Research papers in national conferences. Dr. P. Arjunraj is the resource person for many different Workshops, Training program, Seminar and Guest Lectures in different companies, R&D institutions, Academic Institutions related to Value Engineering, Operation Research, TQM, Manufacturing Technologies, Emission Control Technologies, Electric Vehicle Technologies, Quality, Automotive Engines, Automotive Transmission, Various Automotive Components, Mechanical Product Design, Management Subjects etc. He is also the Motivational Speaker to the students and Faculties in Various Engineering Colleges and different Business Schools. Dr. P. Arjunraj is a

- Life Member in Quality Circle Forum of India (QCFI)

- Life Member in Indian Value Engineering Society (INVEST)

- Life Member in Institution of Engineers, India (IEI)

- Life Member in the International Association of Engineers (IAENG)

- Senior Member of Indian Society of Mechanical Engineers (SMISME)

- Member of Society of Automotive Engineers (SAE)

- Member of Society of American Value Engineering SAVE International, USA

- Education Committee Member in Miles Value Foundation, USA

Dr. P. Arjunraj has given career guidance, motivational speeches and Technical seminar to more than 2 lakh students in various Engineering and Management colleges across India Dr. P. Arjunraj has given awareness about Value Engineering to numerous employees working in many leading organization and companies across various state in India to achieve the concept of "Make in India" through Value Engineering and value methodology. Dr. P. Arjunraj have worked in BEML Limited as an engineer in Service Department, Mahindra and Mahindra Limited in engine research and development, PSA AVTEC powertrain Private Limited in research and development. Dr. P. Arjunraj has a YouTube channel namely Arjunraj P dedicated to engineering and technological topics with easy explanations. Dr. P. Arjunraj has given awareness to numerous professionals to achieve the concept of Make in India through VE/VA. He has received many awards for his outstanding contribution to VE/VA field, such as

- Awarded Professional in Value Management (PVM) by European Governing Board for Value Management and certification for Value Management.

- Recipient of Gold Medal Award 2015 and 2019 by Indian Value Engineering Society for the outstanding contribution related to Value Analysis and Value Engineering in Industry.

- Awarded Miles Value Foundation Paper of the Year 2016 by Society of American Value Engineering.

- South Zone Council Member of Indian Value Engineering Society and Member of SAVE International USA since 2016.

- Recipient of research excellence award 2016 by Indus Foundation during Indo American education Summit

- Served as the champion for engineer new entrepreneur program in society of automatic engineer SAE India Southern section (August 2016 to December 2018)

- Received advance leadership from award from Toastmaster International in March 2017.

- Received Triple Crown award from Toastmaster International on April 2017.

- Recipient of research excellence award 2016 by Indus Foundation during Indo American education Summit.

- Awarded as Chartered Engineer by Institution of Engineering IEI India during July 2017.

- Recipient of Best Student Service award 2017 by Indus Foundation during Endo Global skill summit in July 2017.

- Appointed as management committee member in society of automatic engineer SAE India Southern section Hosur division from December 2018 to November 2019.

- Awarded Distinguished Service in Industry 2021 by SAVE International US USA

- Presidential Citation Honor 2023 award from SAVE International, USA.

Foreword

The automotive industry has undergone significant transformation, shifting from mechanical systems to electronic control units and software-driven functionalities. This shift has brought unprecedented innovation and efficiency, fundamentally altering the driving experience. However, it has also introduced new challenges, particularly in ensuring functional safety.

To address these challenges, the International Organization for Standardization (ISO) has introduced ISO 26262, a crucial standard for automotive functional safety. As vehicles become more advanced, with numerous electronic systems and functionalities, the need for a comprehensive approach to functional safety has become increasingly important.

"Functional Safety in Modern Mobility: ISO 26262 and Beyond," authored by Dr P. Arjunraj, explores this essential topic in depth. The book not only explains the principles and applications of ISO 26262 but also addresses emerging challenges and solutions in the evolving landscape of modern mobility.

Dr Arjunraj's extensive experience and insightful analysis provide a unique perspective on the critical importance of functional safety in today's automotive industry. As vehicle users enter an era of autonomous vehicles and intelligent transportation systems, the insights in this book are more relevant than ever.

"Functional Safety in Modern Mobility: ISO 26262 and Beyond" is a valuable resource for engineers, researchers, policymakers and anyone dedicated to advancing automotive safety.

Dr. P. Mannar Jawahar
Former Vice Chancellor, Anna University,
Former Vice Chancellor, Karunya Deemed University, Coimbatore

Foreword

As the title suggests, this book does not merely confine itself to the existing framework but explores the uncharted territories of the future. In a domain where technology evolves at an unprecedented pace, understanding the nuances of upcoming challenges and opportunities is paramount. The author Dr. P.Arjunraj draws upon his wealth of experience to provide a forward-looking perspective, addressing not only the present needs of the industry but also anticipating the requirements of tomorrow.

As the automotive industry continues its march towards a future marked by autonomous vehicles, electrification, and connectivity, the insights presented in this book will serve as a compass, guiding practitioners and innovators towards safer, more reliable, and technologically advanced transportation solutions. The author Dr. P.Arjunraj has presented the content in a lucid and accessible style, ensuring that even the most intricate concepts are comprehensible to readers with varying levels of expertise.

This book serves as an indispensable guide through the intricate realms of ISO 26262, offering insights, analyses, and practical applications that bridge the gap between theory and implementation. However, its significance extends beyond the confines of this well-established standard. The author Dr.P.Arjunraj knowledge in this field, meticulously traverse the landscape of automotive functional safety, delving into the latest advancements and emerging trends that extend far "beyond" the current standard.

Dr.G.Ranganath
President, The Institution of Engineers (India) (2023-2024)
Principal, Adhiyamaan College of Engineering, Hosur

Acknowledgment

First and foremost, I express my heartfelt gratitude to the divine force that guides and sustains us all, for granting me the wisdom and inspiration to undertake this endeavor.

I want to express my deepest gratitude to the countless individuals who have played significant roles in shaping my life and enabling me to pursue my current path. While the list of deserving thanks is extensive, I would like to acknowledge those who have had the most profound impact on the success of this book. My appreciation extends to the audiences, students, and professionals who have attended my keynotes and training programs over the past few years. Your engagement and feedback have been instrumental in inspiring me to write this book much sooner than I had originally planned. I also want to thank the colleges that invited me to guide their students and the companies that provided me with the opportunity to train their employees in functional safety. Your collective support and enthusiasm have been the driving force behind the creation of this book. Without your encouragement and valuable interactions, this book would not have been possible. Thank you for being the catalysts that propelled me to share my knowledge and insights through this work.

I extend my profound thanks to Dr. P. Mannar Jawahar, Former Vice Chancellor Anna University of Chennai and Former Vice Chancellor of Karunya Deemed University, Coimbatore for his invaluable guidance and mentorship. His wisdom and encouragement have been instrumental in shaping this book, greatly enriching its content.

My sincere appreciation goes to Dr. G. Ranganath, President (2023-2024), The Institution of Engineers (India) and Principal, Adhiyamaan College of Engineering whose unwavering support and insightful feedback have been crucial throughout the development of this book. His dedication to advancing the field of automotive functional safety is truly inspiring.

I am deeply indebted to Ms. Nandini Umapathi, Functional Safety Engineer whose dedication and tireless efforts were pivotal in the research, data accumulation, and coordination with the publishing team. Her unwavering support has been invaluable in bringing this project to fruition.

Special thanks to Mrs. Poonam Patil, for her valuable support during the final stages of this book's development. Her insights and contributions were instrumental in refining the content and ensuring its clarity and coherence.

Lastly, I extend my heartfelt thanks to all those who have been part of this incredible journey, directly or indirectly. Your contributions, big and small, have helped shape this book into what it is today. Thank you all for being a part of this endeavor.

Embark on an enlightening journey into automotive functional safety with this book, a beacon illuminating the path towards a safer and more innovative automotive future.

Chapter 1

Introduction to Functional Safety

Topics to be covered:

1. What is Safety of a function?

2. The difference between functional Safety and Security

3. FuSa standard IEC61508

4. The derivatives of IEC61508

5. Need of Safety in Automotive - ISO26262 Standard

6. Safety systems in Automotive and Scope of ISO26262

7. Objective of ISO26262

8. FuSa as per ISO26262

9. Structure of ISO26262

10. Case study

Safety of a function

Safety of a function is needed when we depend on an automated environment to achieve a task or a function, and the function has to be proven reliable enough not to cause any potential hazard. In other words, the risk is limited to a reasonable level.

Therefore, when we automate a system or an equipment, we need it to respond correctly to all of its inputs and to react predictably in case of failure events. In that way, we will be able to prove the system is reliable even when expected or unexpected failures occur.

Difference between Safety and Security

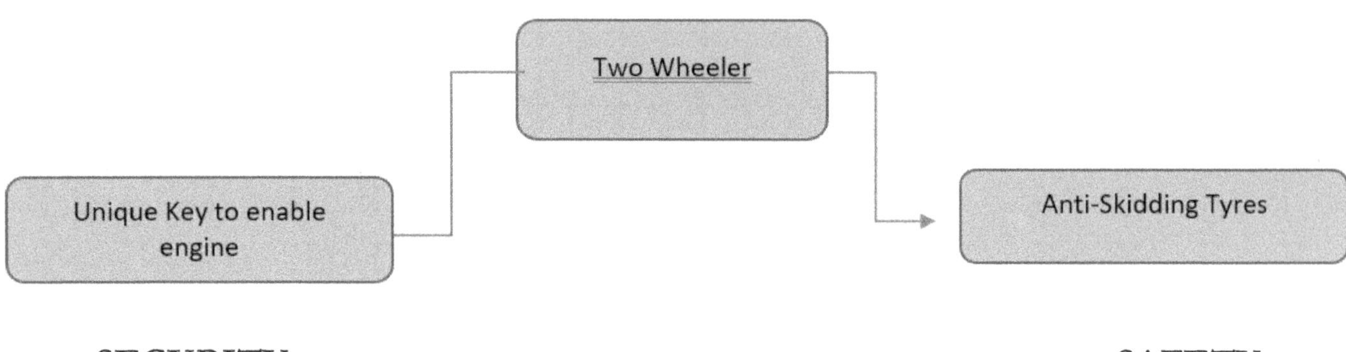

Safety: Safety is defined as avoiding the inevitable risk of personal injury. Safety is the state in which the risk of harm to any person or property is reduced to, and maintained at or below, an acceptable level through a continuing process of hazard identification and risk management. Safety protects the target functional unit from causing danger, risk, or injury to its environment, by mitigating the risks and preventing them by precautionary measures.

Security: Security is the process of protecting the person/ Organization/ building against crime and attack from exterior sources.

Functional Safety Standard IEC61508 [1]

At home, at work or in public places, we are surrounded by more and more electrical and electronic devices and systems. Security efforts focus on electronic equipment and related software and the provision of security procedures to greatly reduce the risk of harming someone or destroying something. The sensors on the automatic doors ensure that the doors open fast enough and close smoothly behind you. Sensors detect smoke and activate the vacuum system in the building. The overflow valve is activated when the liquid or pressure reaches a certain level.

The IEC 61508 series was created by the International Electrotechnical Commission (IEC) to provide safety standards for the life of electrical, electronic or programmable electrical (E/E/PE) systems and products. It affects equipment or system that performs automatic safety, such as sensors, control logic, actuators, and microprocessors.

IEC 61508 allows the development of common standards that can be used for all electrical safety and other software. It is a horizontal model that can be applied to many sectors.

This standard requires an analysis of the risks or hazards of the process or equipment.

It provides categories to determine the level of potential hazard and the consequences of its occurrence. IEC 61508 defines security integration levels (SILs) to indicate the degree to which a system performs its safety function.

The Derivatives of IEC61508

The original IEC 61508 series is an international standard for safety-related technology. The model supports risk assessment to minimize operational failures regardless of where and how the system is used. IEC 61508 is the most important safety standard for all industrial applications, based on two concepts: Safety and SIL. The security lifecycle includes the engineering process that includes all the steps to achieve security. Creates and writes a security plan, then executes it.

On the other hand, there are four SIL levels that calculate the level of risk reduction; SIL 1 is the lowest level of risk reduction and SIL 4 is the highest level of risk reduction. SIL certification identifies process hazards, eliminates the risk of failure, and determines whether the product is safe to fail.

From IEC61508, various standards were derived to make them specific to each sector, as mentioned in the above figure. Few of mentionable examples are:

EN50128 for Railway applications

When you pass by the train, the security function can be sounded so that the door closes before the train starts to move and does not open while it is moving. In an emergency, the system alerts trained pilots to take control.

ISO26262 for Automotive - Passenger vehicles up to 3500kg

In our cars, the safety function ensures that the airbags are only deployed immediately in the event of a collision, not while driving. In addition, fuel injection control ensures that the vehicle only accelerates when commanded. The brake system is activated when necessary. In today's cars, safety functions ensure the correct operation of all the car's electronics as well as the control software.

ISO25119 for Machinery used for Agriculture and Forestry

In agriculture and forestry, safety studies are carried out to ensure that machines are equipped with protection against special soils and heavy use. For example, tractors are equipped with suitable tires and tires and safety measures to prevent accidents. It is also suitable for transporting mobile trucks, we move the box up and down for loading and unloading. The standard ensures that such vehicles are equipped with appropriate warnings and error detection to minimize hazards.

Need of Automated Safety in Automotive

In February 2009, a Swiss insurance company concluded that 90% of rearend collisions could be avoided with a 1.5-second warning. Also, the 2.0 second warning can avoid almost any collision. [2]

Scientists have given us the term "recognized response time (PR time)"; this means the time it takes for the driver to first see a problem from the moment they apply the brakes.

A University of Michigan study concluded that 95% of drivers experienced a PR time of about 1.6 seconds when faced with the blind spot. Currently this PR decision is based on psychological data and may be higher considering the environment. This long process consists of 3 parts: the time it takes for the pilot to perceive the obstacle, determine the appropriate decision, and act accordingly.

According to American practice, the available PR time is about 2.5 seconds, including fatigue and environment. [3] This has been removed by security management. These machines can calculate and provide millisecond-accurate collisions that are impossible for humans to do.

For all electrical applications, there are safety standards that define the design, construction and management of electrical equipment to minimize the occurrence and consequences of accidents or vehicles or hazards that may result from electrical equipment failure. For the automotive industry, ISO26262 is the standard that provides us with the rules and regulations for establishing and maintaining safety procedures.

Types of Safety Systems in Automotive

Active Safety Systems: The active safety systems are designed to mitigate and prevent accidents and hazards. Few examples presented in the figure.

Passive Safety Systems: These systems respond when an accident occurs. It is designed to cushion and protect the driver and passengers in the vehicle to prevent accidents.

A simple difference between passive safety systems and active safety systems in the automobile industry is that the former reduces the likelihood of accidents and ultimately prevents them.

Scope of ISO26262

The Scope of ISO26262 is the Active Safety Systems.

Functional safety, as per ISO26262, is the absence of unreasonable risk due to hazards caused by malfunctioning behavior of Electrical/Electronic (E/E) and software systems.

The ISO26262 reduces the risk of any E/E system in the automotive by:

- Identifying Hazard (by performing Risk Analysis of the system)

- Measuring associated risk (by defining ASIL levels for identified risks)

- Lower the risk to an acceptable or reasonable risk (by integrating safety mechanisms)

The ISO 26262 standard addresses the hazards that may arise from the failure of safety-related E/E systems, including the intervention of these systems. It does not address issues related to electric shock, fire, fire, fire, chemical, fire, fire, fire electrical, electrical discharge, and similar issues, except for the failure of direct E/E systems.

These comply with the EMC and performance requirements specified by the vehicle manufacturer and are recognized by the specific regulations (national laws) of the end user location. Non-E/E systems are covered by ISO13849, another standard also provided by IEC61508.

The Release of ISO26262 Standard:

ISO (the international organization for Standardization) is a global federation of countrywide standards bodies (ISO member bodies). The paintings of making ready worldwide standards is commonly achieved through ISO technical committees.

There are revisions of ISO26262 released by above noted committee:

- ISO 26262:2011

- ISO 26262:2018 - the latest and revised standard version

This edition of ISO 26262:2018 collection of standards cancels and replaces the edition ISO 26262:2011 collection of requirements, which has been technically revised and includes the subsequent main adjustments:

- necessities for vehicles, buses, trailers and semi-trailers

- goal oriented confirmation measures

- management of protection anomalies

- references to cyber protection

- up to date target values for hardware architecture metrics

- steering on version-primarily based development and software safety analysis

- assessment of hardware factors

- additional steerage on structured failure evaluation

- guidance on fault tolerance, safety related special traits and software tools

- steerage for semiconductors

- necessities for bikes; and

- general restructuring of all elements for improved readability.

Objective of ISO26262 standard

To achieve functional safety, the ISO 26262 series of standards:

1. Provide information for the traffic safety cycle and support activities that must be carried out in the life cycle (for example, development, production, operation, service and decommissioning).

2. provides a risk-based vehicle specific method to determine the integrity level [Automotive Safety Integrity Level (ASIL)]

3. uses ASIL to determine the ISO 26262 requirements used to prevent residual risk

4. provides arrangements for security management, design, use, validation, validation and evaluation of validity;

5. specifies the requirements for the relationship between customers and suppliers.

FuSa as per ISO26262 standard

Safe operation is the absence of unnecessary risks due to hazards arising from bad behavior of electrical/electrical systems and software.

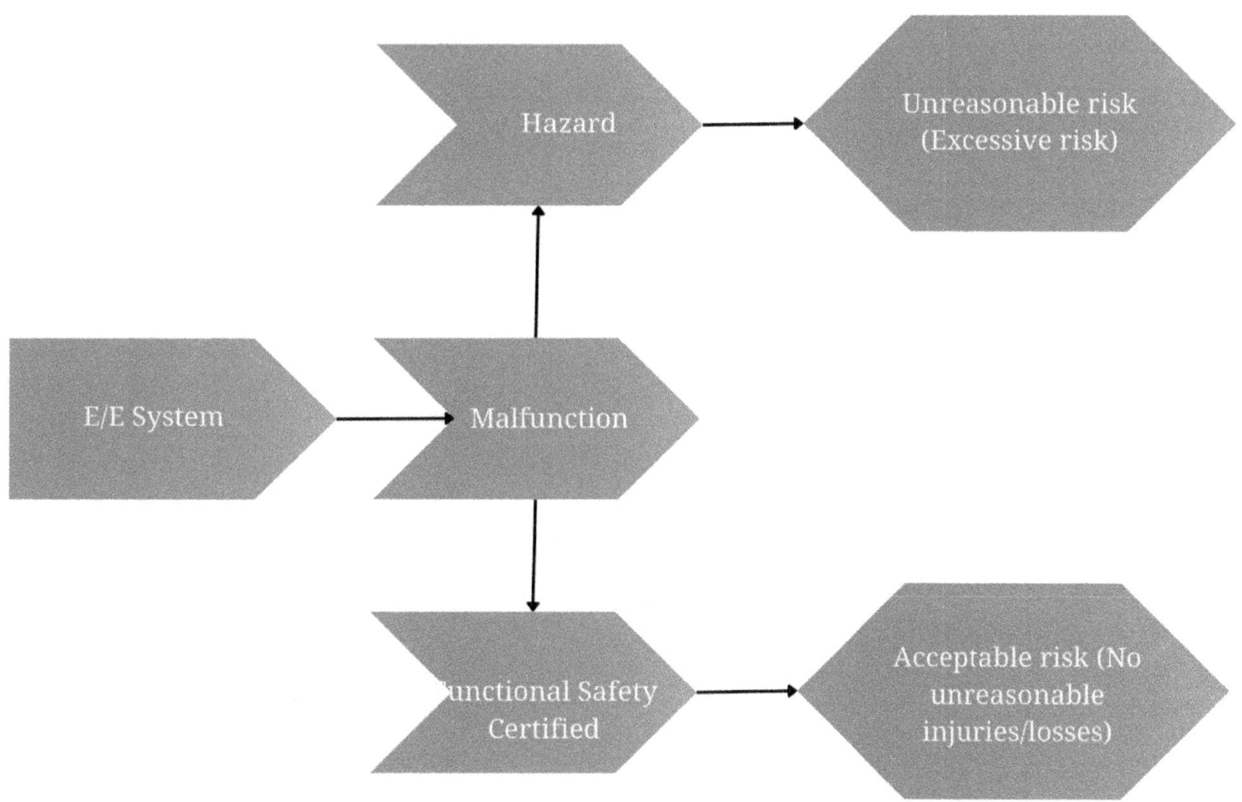

Example System:

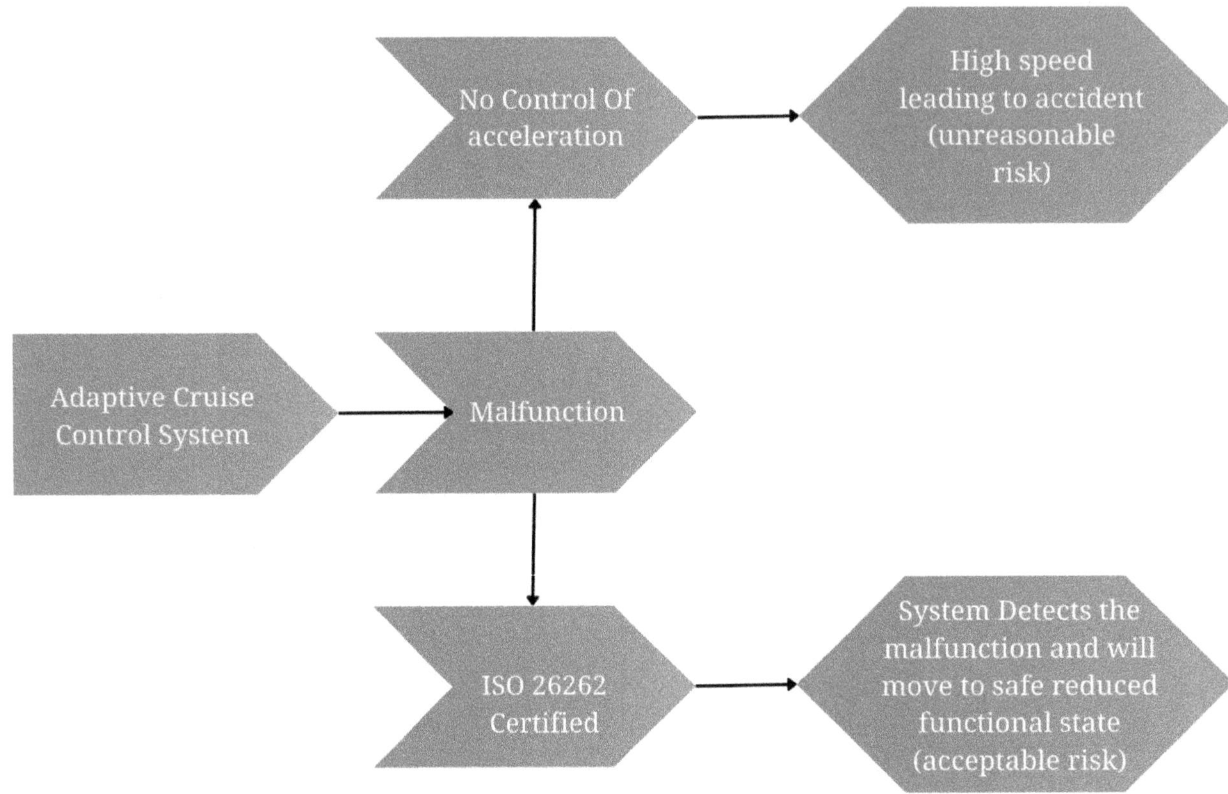

ISO 26262 specifies the requirements for safety related to the operation of the system and to meet the procedures, processes and equipment used during construction. The ISO 26262 standard provides adequate safety and control over the life of the vehicle.

ISO26262 Structure

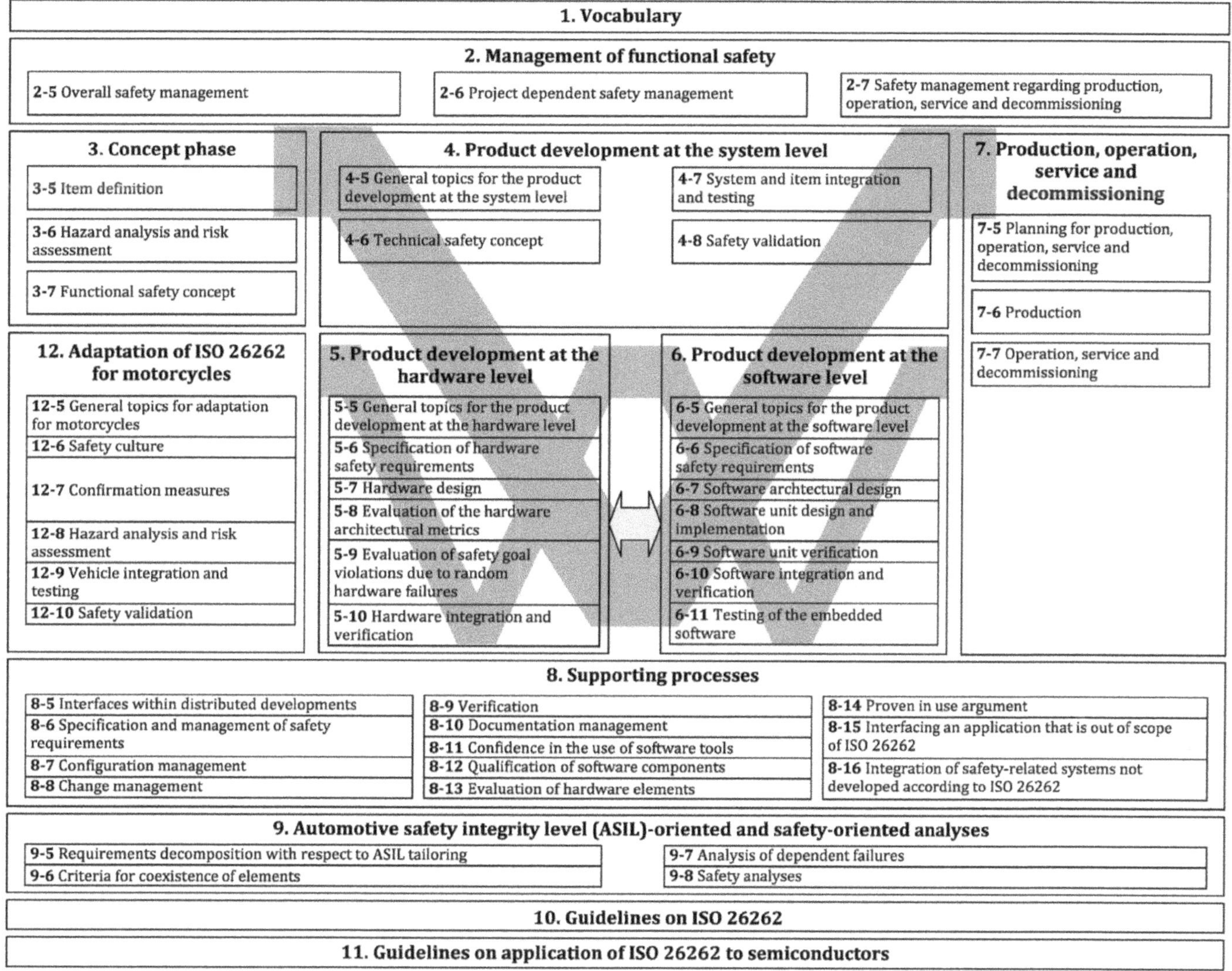

For development of E/E systems in automotive industry, this standard covers the rules and requirements for the complete product development process and the steps are defined to be as below:

1. Requirement Specification

2. Design

3. Implementation

4. Integration

5. Verification

6. Validation

ISO 26262 is an international standard for the functional safety of electrical and electronic systems in automobiles. It provides guidelines for the development, implementation, and management of safety-related systems throughout the automotive product development lifecycle. The standard is organized into 12 parts, each addressing specific aspects of functional safety. Here is an overview of these parts:

Part 1: Vocabulary

Part 1 of the standard is the "Dictionary of the ISO". It outlines the terms and abbreviations used by the standard. This part provides definitions of key terms and concepts used throughout the standard. It establishes a common understanding of the terminology related to functional safety.

Example:

ASIL – Automotive Safety Integrity Level

E/E System – Electrical / Electronic System

Part 2: Management of Functional Safety

This part focuses on the management aspects of functional safety. It outlines the requirements for the management system, safety management, and safety culture within an organization. It provides the ISO26262 requirements in 3 sets of clauses.

- The first of these relates to the development organization.

- The second is applicable to each development projects and

- The third relates to the post-development phase.

There are other clauses of ISO 26262 Part 2 and they define further requirements that are closely related to managing functional safety. For functional safety in automotive, Part 2 defines the Safety development life cycle of the project. Project specific life cycle needs to be defined in the Safety Development Plan. As mentioned above, the Part 2 provides the guidelines for the Management of functional safety that needs to be addressed

- on an organizational level,

- on a project level and

- for the time after release for production.

Part 3: Concept Phase

ISO 26262 refers to the early phase of product development as the "concept phase", and mainly describes it in part 3. This part provides guidelines for the safety activities during the concept phase of the product development lifecycle. It covers the identification of safety goals and requirements, hazard analysis, and the creation of a safety concept. The Concept phase also includes an impact analysis required in Chapter 2. For use in motorcycle design, Chapter 12 provides more specific information on risk assessment.

In the concept phase, the following four topics have to be implemented.

1. Item definition - The subject of development, the item, must be defined and its boundaries determined.

2. Impact analysis at the item level - If the target product is a modification of previous product or it to be used in a different environment, Impact analysis will be done to determine necessary safety activities to be performed.

3. Hazard analysis and risk assessment - The item is subjected to a hazard analysis and risk assessment, in order to scale safety activities.Safety goals are formulated.

4. Functional safety concept - The functional safety requirements are assigned to systems for implementation. A functional safety concept describes, in a panoramic way, how the hazards should be mitigated.

Part 4: Product level at the System level

The fourth part deals with development processes on a system level in keeping with the V-Model. It addresses the safety activities during the system development phase. This includes system-level hazard analysis, functional safety concept, and technical safety requirements. The system level follows the concept phase in the life cycle.

Functional safety requirements are to be detailed into technical safety requirements so that they can be implemented with the system architecture to be developed in parallel. The Systematic failures and random hardware failures need to be identified and addressed at this level. This is done with the aim of securing the specification of safety requirements, safety mechanisms, design and validation.

Part 5: Product level at the Hardware level

The fifth part focuses on the safety activities related to hardware development. It covers the hardware requirements, design, and verification for achieving functional safety. It deals with development processes on a hardware level in keeping with the lower segment of the V-Model. Hardware development is part of system development in the safety lifecycle and runs corresponding to software development.

In this part of the development cycle, the security system must define good hardware requirements for use in hardware development. Hardware failures should be classified according to how they directly violate security objectives. Evidence must be provided to show that the equipment malfunction did not violate the safety purpose and did not remain undetected in the vehicle. In addition, the system should be used to model hardware. And it must complete hardware testing according to industry standards.

Part 6: Product level at the Software level

Part 6 is devoted to the development of the software level following the subsection of the V-pattern. It provides guidelines for the software development process in relation to functional safety. It covers software requirements, architectural design, and verification activities.

With the advancement of software in cars and the further progress of driving, the safety of cars is increasing on the basis of error-free software. Software error should also be prevented by improving performance. Failures must be handled through fault-tolerant mechanisms. The security system of the system should be optimized for the security of the software. Similar to Chapter V, software integration and testing must be implemented in various ways, and of course testing must be done before final software is released.

Part 7: Production, Operation, Service, and Decommissioning

Part 7 addresses the safety activities during the production, operation, service, and decommissioning phases of the product lifecycle. It includes guidelines for manufacturing, operation, maintenance, and eventual decommissioning of the system. It deals with the production and planning process. The aim is to meet safety requirements during manufacture and installation. The rules and recommendations of this article apply to production planning, operation, service and removal of products and equipment, including their installation in vehicles.This clause includes requirements to ensure that functional safety is achieved during the production process by including these safety-related special characteristics in production planning and control. This clause also includes requirements related to developing service information and user information (including the user manual); the planning, execution and monitoring of maintenance work; repair instructions; decommissioning instructions; and instructions and information for rescue services.

Part 8: Supporting Process

The aim of the eighth part is to define and delegate responsibilities. It covers supporting processes that are crucial for achieving functional safety, such as configuration management, change management, and documentation. The customer (e.g. vehicle manufacturer) and the suppliers for item or element developments jointly comply with the requirements specified in the ISO 26262 series of standards for distributed developments.

Responsibilities are agreed between the customer and the suppliers for the concept, development, production, operation, service and decommissioning phases of the safety lifecycle.

Subcontractor relationships are permitted. The customer has safety-related procedures concerning planning, execution and documentation for in-house item developments, therefore comparable procedures apply for co-operation with the supplier on distributed item developments. The same applies for item developments where the supplier has the full responsibility for functional safety.

The requirements of this safety life cycle are specified and configuration and change management are explained in this clause. This also involves defining how tools are used.

Part 9: ASIL- Oriented and Safety-Oriented analyses

This part 9 provides detailed guidance on the safety analysis techniques, such as hazard analysis and risk assessment, to determine the Automotive Safety Integrity Level (ASIL) for the system defines the guidance to reach the ASIL of the safety goals of an item under development.

The safety requirements are allocated to architectural elements, starting with functional safety requirements allocated to elements of system architectural design and finally resulting in safety requirements allocated to the hardware and/or software elements. This part defines the process of how to confirm the ASIL met for the implemented Safety requirements.

It also covers the ASIL decomposition, which is a method of ASIL tailoring during the concept and development phases. During the safety requirements allocation process, benefit is to be obtained from architectural decisions including the existence of sufficiently independent architectural elements.

Part 10: Guideline on ISO 26262

The tenth part offers additional guidance on the interpretation and application of ISO 26262. It provides examples, explanations, and clarifications to assist organizations in understanding and implementing the standard. It provides examples of applications and supplementary details on ISO 26262. This part is more for information purposes.

Part 10 of ISO26262 describes more about the safety management, guidance on work products for concept phase, system development, Safety process requirement structure, concerning Hardware development, example of proven in use arguments and discussion on emergency operation time interval and fault tolerant systems.

Part 11: Guideline on Application of ISO 26262 to semiconductors

This part focuses specifically on the application of ISO 26262 to the development of semiconductors used in automotive systems. It provides guidance on semiconductor-specific safety requirements and considerations. A semiconductor component can be developed in two ways - System Element in Context and System Element out of Context(SEooC). In the SEooC method, the development is done based on assumptions on the conditions of the semiconductor component usage, and then the assumptions are verified at the next higher level of integration considering the semiconductor component requirements derived from the safety goals of the item in which the semiconductor component is to be used. Part 11 of ISO26262 deals with the descriptions and methods that are to be implemented when we develop a semiconductor component as a SEooC.

Part 12: Adaptation of ISO 26262 for Motorcycles

Part 12 addresses the adaptation of ISO 26262 for motorcycles, taking into account the unique characteristics and safety considerations specific to motorcycles. It places more responsibility on the motorcyclist rather than the motorcycle to mitigate risks. Motorcycles were not explicitly excluded in the standard, the standard was essentially applicable only on cars or passenger vehicles.

ISO26262 defines Part 12 especially for Two wheelers. For items or elements of motorcycles for which requirements of ISO26262 are applicable, the requirements of this part 12 supersede the corresponding requirements in other parts.

Case Study

Functional Safety in Instrument Cluster [6]

Let's take the example of a dashboard in a car. The fuel pump of the vehicle has many indicators and signals that enable the driver to start the vehicle. These include various gauges (mostly speedometer, odometer, tachometer, fuel gauge, fuel gauge, etc.) and various indicators for faults and warnings. The instrument cluster provides the driver with a central and easy-to-see location for all important information.

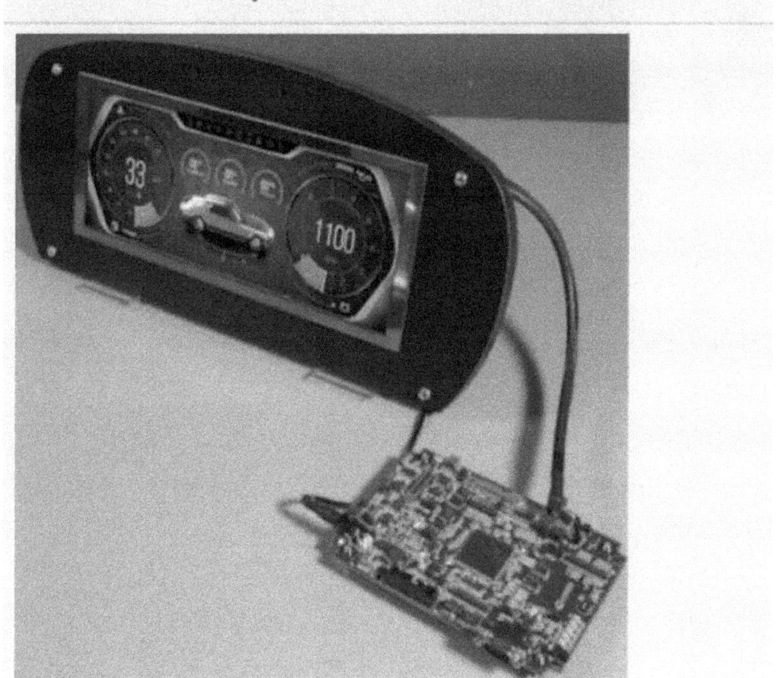

The major subsystems would include:

1. LED load drivers

2. Connectivity interfaces

3. Processor and Memory

4. Power Management

To mention an example, let's take the memory cell as an example. The characters or characters displayed in the TFT display group will be pre-recorded data stored in the memory unit. In Telltale (in our case the dipped beam indicator must be on), the processor orders the display of predefined characters in the address.

Let us now consider an error case where the flash memory stores corrupted low-level characters. The output of the bad memory will represent the first image below. This happens when we have no idea to confirm the accuracy and completeness of the information.

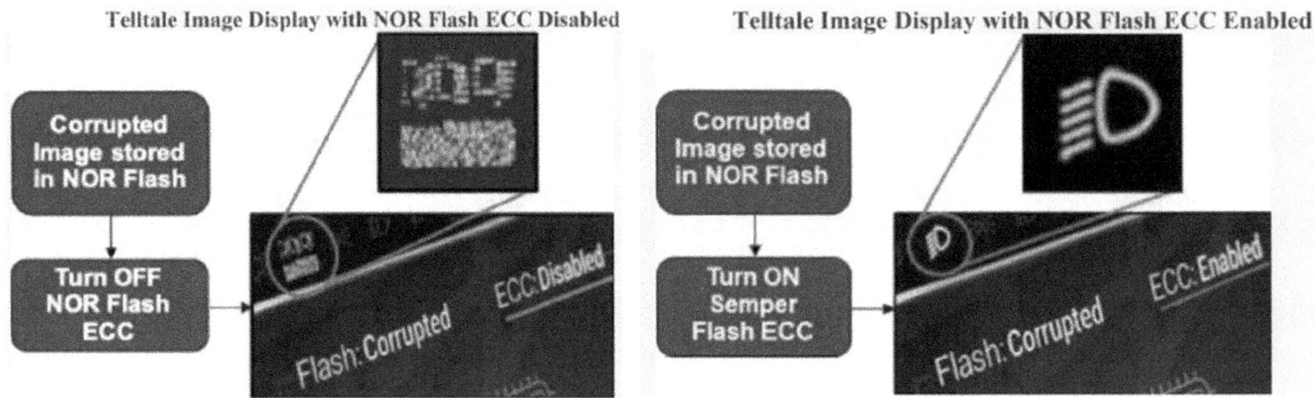

Let us say, we update the design to follow ISO26262 rules and implement ECC safety mechanism in the memory. The ECC mechanism will detect and correct the erroneous bit-flip, hence the low-beam symbol will be correctly displayed. Few more examples that can be implemented for a memory unit are CRC check, redundant storage, etc. The above was an example for a memory unit, similarly when we follow the ISO26262 standard right from concept phase and throughout the System, Hardware and Software design and validation, the System will have protection and safety mechanism for almost most of the predictable failure cases of the system. The list of Safety mechanisms that can be defined for various blocks of an instrument cluster is discussed in the below table:

Table 1. Functional Safety in the instrument cluster Traveo II MCU

MCU Elements	Safety Target	Safety Mechanism	Function Description
CPU	Processing, Bus Communication	Self-Test Library	CPU Test Software
		Window Watchdog Timer	Timeout Monitor
Embedded Flash Memory	Memory Cell, Data	ECC	2-bit Detection and 1-bit Correction
Embedded RAM	Memory Cell, Data	Self-Test Library	Initial Testing by Software
		ECC	2-bit Detection and 1-bit Correction
ADC	Analog Connection and Conversion	Analog Diagnostics	Reference Conversion
Communication Peripheral	Data	CRC Unit	Checksum Calculation
Graphics	Display Output	Signature Unit	Checksum Calculation
Power Supply	Voltage Monitor	Low-voltage Detection	Error Monitor for Voltage
Clock	RC/Crystal Oscillator/PLL	Clock Supervisor	Clock Lost Detection, Frequency Anomaly Detection
Software	Software Partitioning	Memory and Peripheral Protection	Access Protection
	Program Flow	Timing Protection	Time-Out Monitor

There are many other components and systems that are automated in a car. ISO26262 shall be followed to meet the requirements to ensure safety of such automated systems. Below image has some examples of such systems, with corresponding possible failure modes and ASIL levels.

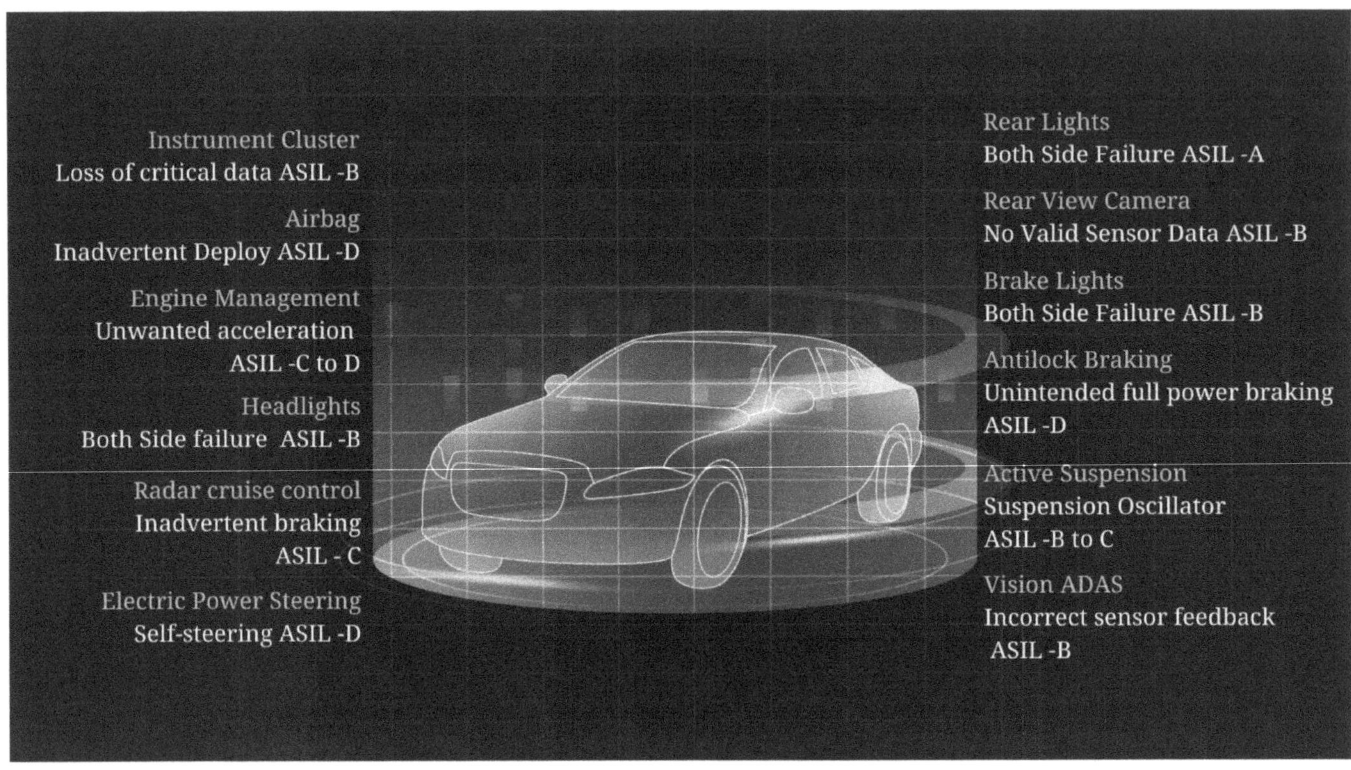

Chapter 2

ISO26262 PART 1 – VOCABULARY

Topics to be covered:

1. Purpose of Part 1 of ISO26262

2. Important terms in ISO26262

3. Terms for classifying vehicles

4. Terms for Levels of architecture

5. Terms for Safety Requirements

6. Terms used in association with Safety Requirements

7. Terms for Faults and Failures

8. Systematic vs random faults/failures

9. Dependent vs independent failures

10. Types of dependent failures - Common cause vs Cascading failures

11. Terms for ISO26262 metrics

12. Important Abbreviations in ISO 26262

Purpose of Part 1 of ISO26262

The purpose of Part 1 of ISO26262[1] is to define the vocabulary of terms used in the ISO 26262 series of standards. This clause provides explanation on the abbreviations and terms used in the further part of the standard. This part is the glossary of ISO26262 standard.

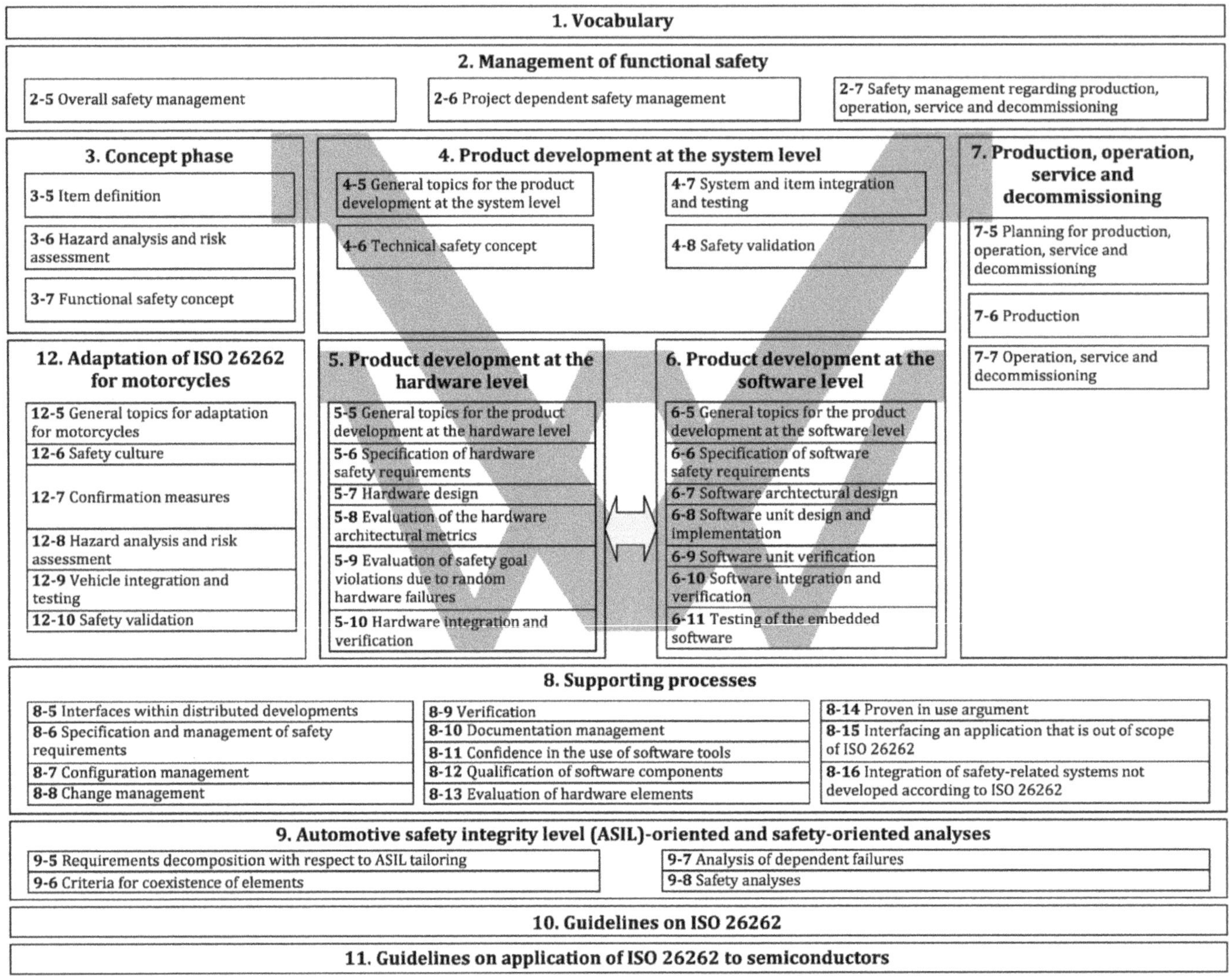

Important terms in ISO26262

Terms for classifying vehicles

Passenger Car – vehicle designed and constructed primarily for the carriage of persons and their luggage, their goods, or both, having not more than a seating capacity of eight, in addition to the driver, and without space for standing passengers. ISO26262 deals with passenger cars with maximum gross weight of 3500 kg.

Motorcycle – two-wheeled motor-driven vehicle, or three-wheeled motor-driven vehicle whose unladen weight does not exceed 800 kg, excluding mopeds as defined in ISO 3833.

Truck – motor vehicle designed to transport goods, or equipment on-board the chassis

Bus – motor vehicle which, because of its design and appointments, is intended for carrying persons and luggage, and which has more than nine seating places, including the driving seat. A bus may have one or two decks and may also tow a trailer.

Trailer – road vehicle which is designed to be towed such that no substantial part of the total weight is supported by the towing vehicle. A trailer can be designed to transport goods, equipment, or people.

Semi-Trailer – trailer that is designed to be towed by means of a kingpin coupled to a tractor that imposes a substantial vertical load on the towing vehicle

Terms for Levels of architecture

Item – system or combination of systems, to which ISO 26262 is applied, that implements a function or part of a function at the vehicle level.Example: The Airbag deployment function involves the crash detection system, airbag deployment system. Items have Safety Goals with assigned ASILs.

System – set of components or subsystems that relates at least a sensor, a controller and an actuator with one another Example: The Crash detection system detects the crash and informs the Body control Unit. This system consists of the sensor, controller and the output unit that transmits the information.

Element – A collection of ingredients that make up a system or part of a system. It contains the Hardware and Software components. Example: Ultrasonic Sensor Element that provides true/false output in specific interface

Component – logically and technically separable part made up of one or more hardware parts and one or more software units. Example: Ultrasonic Sensor Component, Signal processing unit, Signal processing algorithm that decides true/false, transmitter unit that modulates the signal to be compliant with output interface

Hardware part – a piece hardware which cannot be subdivided. Example: Resistor, Capacitor, Power Regulator IC, Microcontroller IC, etc.

Terms for Safety Requirements of an item/element

Automotive Safety Integrity Level (ASIL) – The ISO26262 standard provides an automotive-specific risk-based approach to determine integrity levels [Automotive Safety Integrity Levels (ASILs)] for any equipment or system. It is notionally corresponding to an IEC 61508 term

SIL (Safety Integrity Level), but not equivalent. This can be applied to a Safety Goal of the item/element.

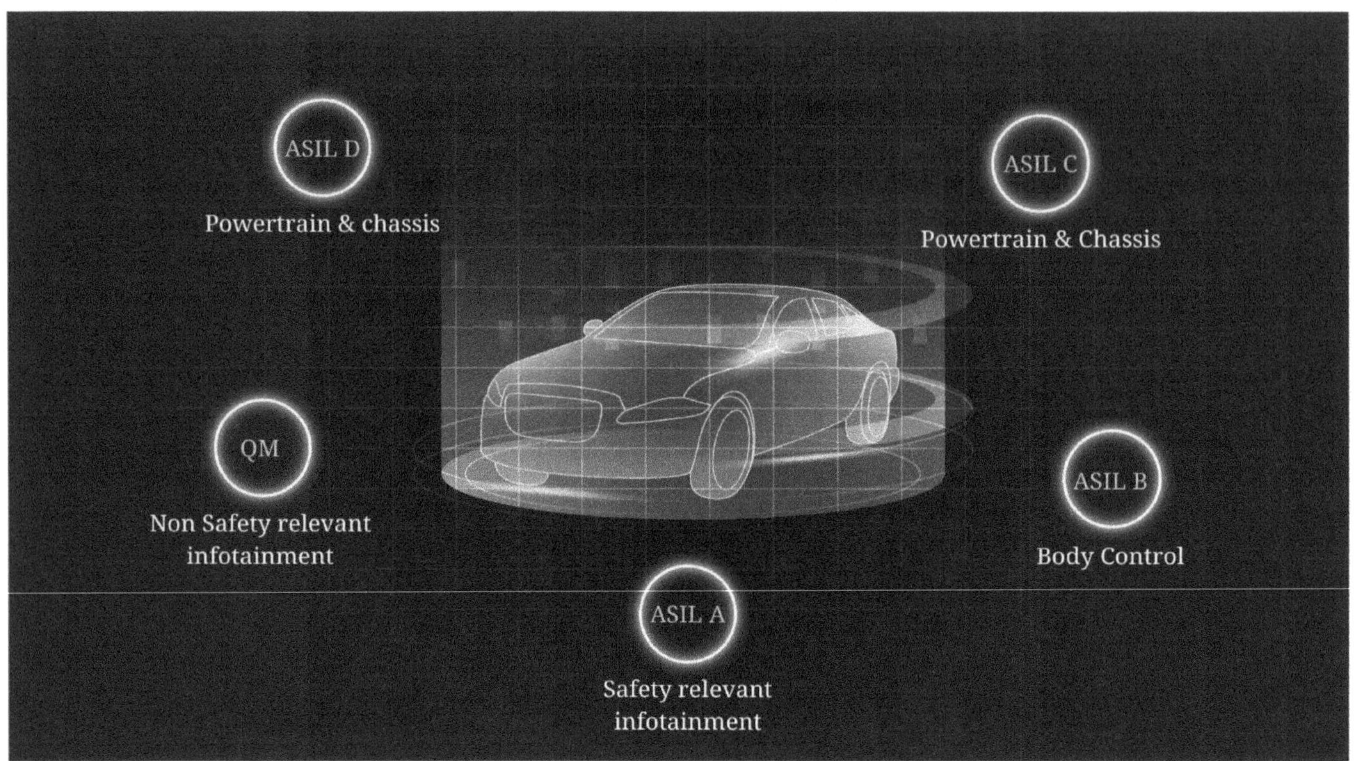

ISO26262 uses one of four ASIL levels (A,B,C,D) to specify the item's or element's necessary ISO 26262 requirements and safety measures to apply for avoiding an unreasonable risk, with D representing the most stringent and A the least stringent level. There is one more rating, QM. This rating is not an ASIL but it states that the requirement can be satisfied by Quality Management and does not require the item/element to be compliant with ISO26262 standard and this is proved during the risk-based approach analysis.

Motorcycle Safety Integrity Level (MSIL) – The definition is similar to ASIL. This term is followed for Motorcycles.

Hazard and Risk Assessment (HARA) – HARA is a method used to identify and categorize hazardous events of items and to specify safety goals and ASILs related to the prevention or mitigation of the associated hazards in order to avoid unreasonable risk. The categorization during this risk-based approach analysis is done by determining ASIL. Tthe ASIL for the specific Safety requirement is determined by three factors - Severity, Exposure and Controllability.

Hazard – potential source of harm caused by malfunctioning behavior of the item/element.

Unreasonable risk – risk judged to be unacceptable in a certain context according to valid societal moral concepts

Severity – estimate of the extent of harm (3.74) to one or more individuals that can occur in a potentially hazardous event

Exposure – state of being in an operational situation (3.104) that can be hazardous if coincident with the failure mode under analysis

Controllability – ability to dodge a specified harm (3.74) or damage through the timely reactions of the persons involved, possibly with support from external measures

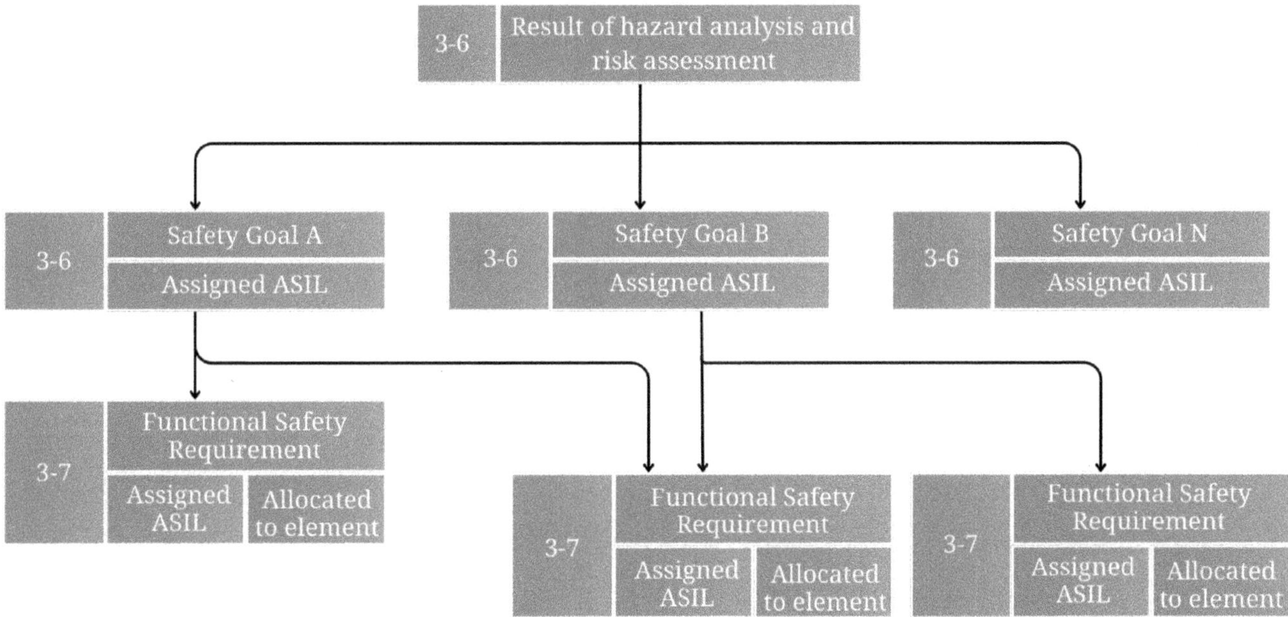

As per above image from ISO26262 Part 3[2], The HARA will result in multiple safety goals corresponding to the different risks identified. Each Safety goal will require multiple Functional Safety Requirements to meet the defined ASIL. Similarly, each Functional Safety Requirement will require one or more than one Technical Safety Requirement to meet the ISO26262 requirements for the specified ASIL.

Safety Goal – top-level safety requirement that is a result of the hazard analysis and risk assessment at the vehicle level. Example: Safety Goal for Steer-by-wire system: The Steer-by-wire system shall not provide erroneous steering output. (ASIL B)

Functional Safety Concept – document containing of the functional safety requirements, with associated information, their allocation to elements within the architecture, and their interaction necessary to achieve the safety goals

Functional Safety Requirement – Necessary safety measures defined as individual requirements at System level including its safety-related attributes. Safety-related attributes must include information about the ASIL and the element to which the requirement is associated.

A functional safety requirement can be a safety requirement implemented by the target system, in order to achieve or maintain a safe state for the item to prevent the hazardous event determined in HARA. Example: The Steering Angle Sensor(SAS) shall implement error-check mechanism to prevent erroneous input to system (ASIL B) (Assigned to Sensor element)

Technical safety concept – document containing the technical safety requirements and their allocation to system elements.

Technical safety requirement – requirement derived for implementation of associated functional safety requirements. The derived requirement includes requirements for fault mitigation and prevention by defining Safety Mechanisms. Example: The Microcontroller shall implement CRC mechanism for the output message "SAS value" to detect error in message transmission. (ASIL B) (Assigned to Microcontroller in SAS sensor)

Safety Mechanism – technical solution implemented by the individual elements within the target system to detect and mitigate, or to tolerate faults, or to control and avoid failures in order to maintain intended functionality in case the fault can be tolerated, or to achieve and maintain a safe state in case the fault will impact proper functionality of the system.

The safety mechanism is either:

a. able to transition to and maintain the item in a safe state, or

b. able to alert the driver such that the driver is expected to control the effect of the failure

Example: CRC mechanism, Timeout mechanism, Frame counter addition are some examples of Safety mechanisms that can be implemented for communication.

Safe State – operating mode without an unreasonable level of risk, defined for the system to reach in event of a non-tolerable failure. While normal operation can be considered safe, the definition of safe state is only in the case of failure in the context of the ISO 26262 series of standards. Example: Warn driver and deactivate engine incase of Safety Goal violation detected in the Steer-By-Wire System

ASIL capable Item/Element – capability of the item or element to meet assumed safety requirements assigned with a given ASIL. Example: If specified the element, say a Microcontroller is specified to be capable of ASIL B, it means that the design of the microcontroller satisfies all the requirements of ISO26262 by implementing the necessary safety measures and meeting the ASIL B metrics. It also means that the microcontroller has the Evidence of Safety implementation meeting ASIL B.

Terms used in association with Safety requirements

Safety Relevant Time Intervals

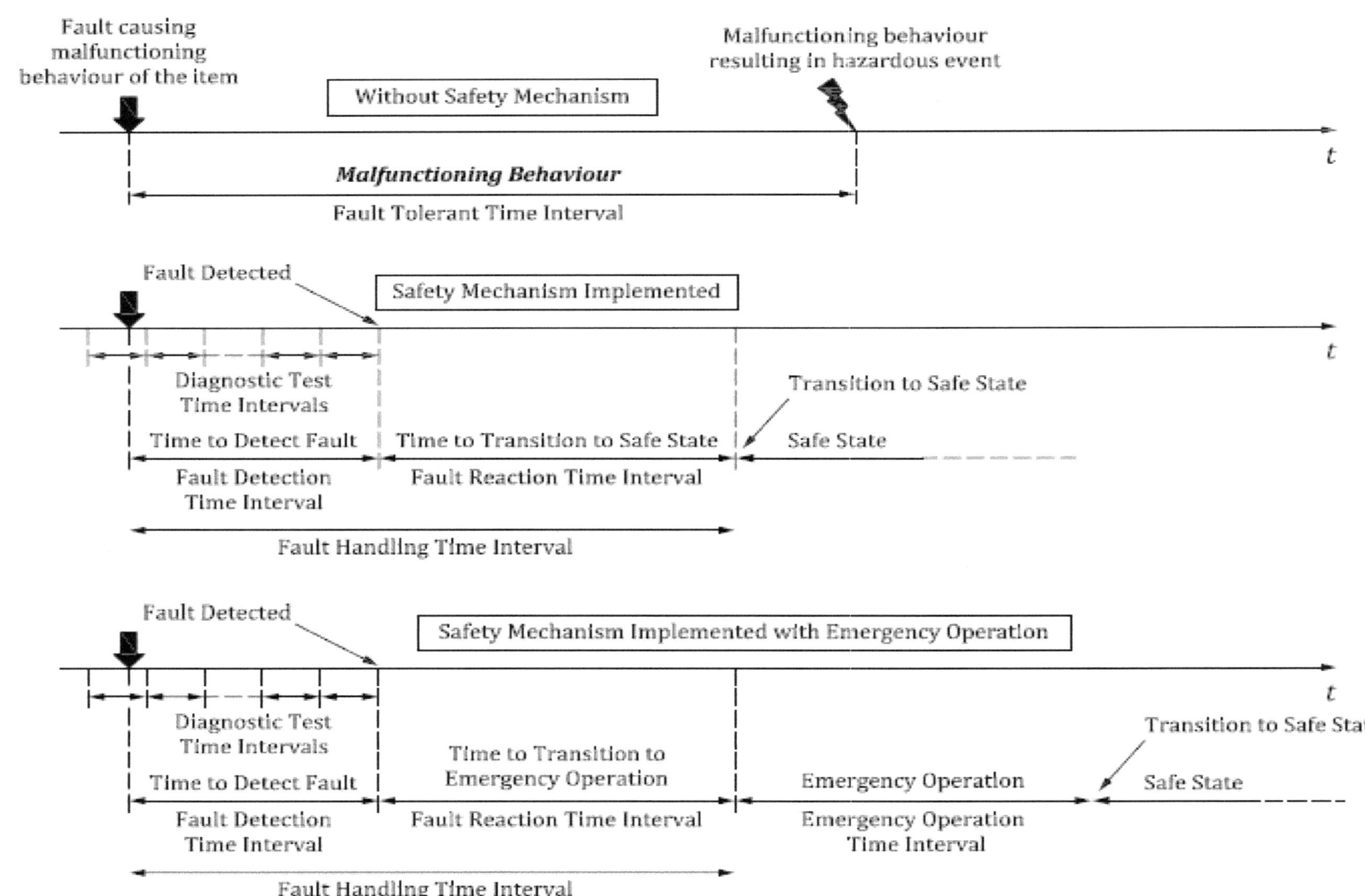

Figure 5 — Safety relevant time intervals

Fault Tolerance Time Interval FTTI – minimum time-span from the occurrence of a fault in an item to a possible occurrence of a hazardous event, if the safety mechanisms are not activated. Example: A failure in the brake system may not result in a hazardous event until the brakes are applied.

Fault Reaction Time Interval FRTI – time-span from the detection of a fault to reaching a safe state or to reaching emergency operation

Fault Handling Time Interval FHTI – sum of fault detection time interval (FDT) and the fault reaction time interval (FRTI). The FHTI is a property of a safety mechanism and to be defined for each technical safety requirement. Example: Let us say, the Steer-by-wire system in event of failure will lead to hazard in 100ms. The 100ms value would be the FTTI of the system, the sum of FDT and FRT (FDTI+FRTI = FHTI) should be within FTTI for the system to reach a safe state before causing a hazard. The Safety requirement will be defined as "The Microcontroller shall check the CRC of the "SAS value"message every 40ms. Incase CRC check fails, the microcontroller shall reach safe state within 30ms." Here, the Fault detection time is 40ms and the fault reaction time is 30ms. Their sum FHTI is 70ms, which is within the 100ms FTTI defined.

Emergency Operation Tolerance Time Interval EOTTI – specified time-span during which emergency operation can be conserved without an unreasonable level of risk. Emergency operation tolerance time interval is the maximum value of the emergency operation time interval. Emergency operation can be considered safe due to the limited operation time as defined in the emergency operation tolerance interval.

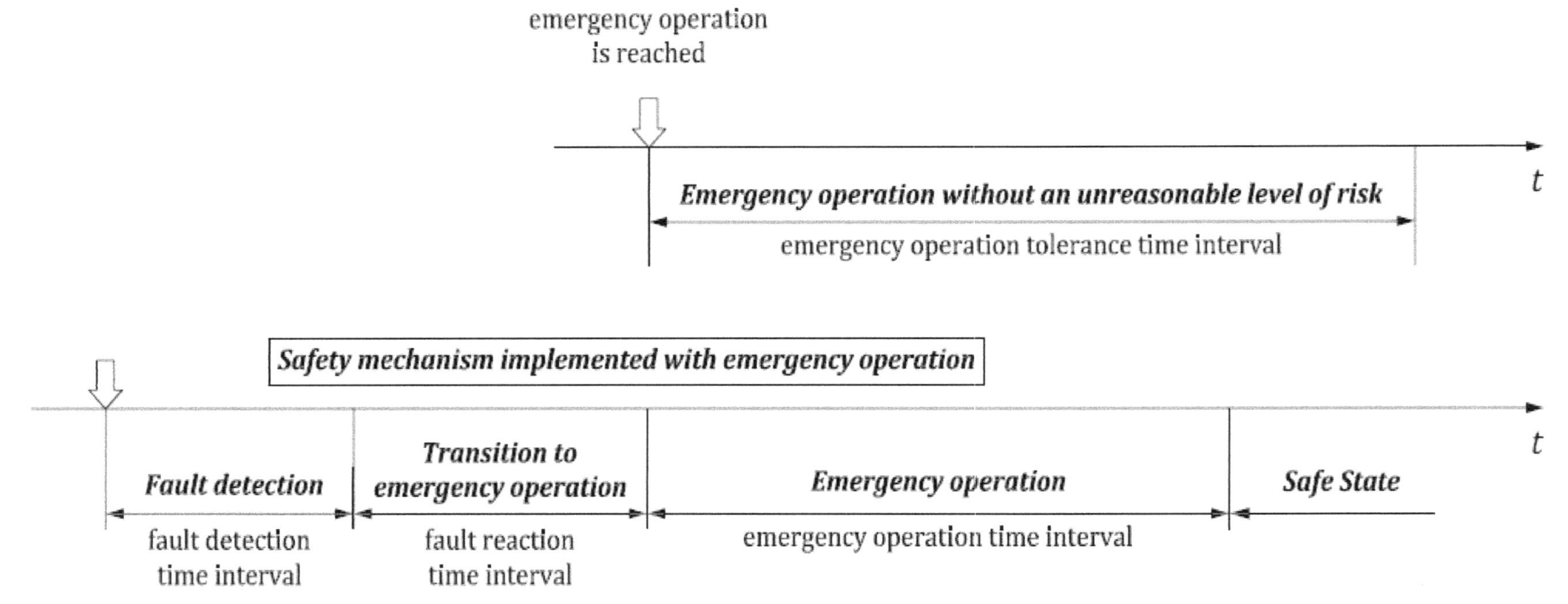

Figure 4 — Emergency operation tolerance time interval

Emergency Operation – When a safe state cannot be directly reached, or cannot be timely reached, or cannot be maintained after the detection of a fault, a safety mechanism can transition the item to emergency operation for providing safety until the transition to a safe state is achieved and maintained. Example: An ASIL B RADAR microcontroller from Texas Instruments AWR1843 has limp mode. This mode is activated whenever Clock Error is detected. If the main clock is detected to be erroneous, instead of completely shutting down by entering reset mode, this microcontroller switches to the limp mode where it works at low frequency, supplied by an independent low frequency clock source. In this mode, the Microcontroller cannot perform all the functions, but it can support to reach a safe state of the system. The microcontroller is allowed to function in limp mode for EOTTI.

Terms for Faults and Failures

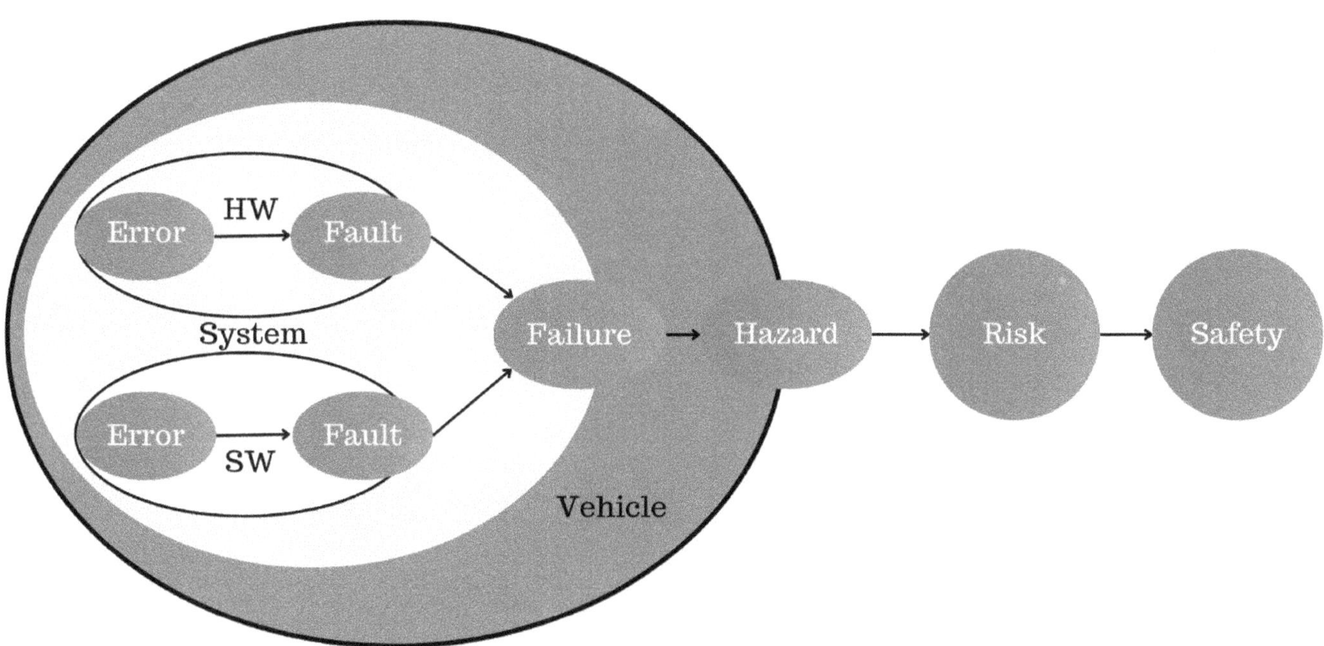

Fault – a fault refers to an abnormality or defect in a system or process that deviates from its intended or expected behavior that can cause an element or an item to fail. The fault is an abnormality that happens at a HW part or Software component level which leads to failure at Item or Element level. In ISO26262 context, failure means violation of safety goal.

Types of Faults: [4]

a. Perceived Fault – detected by the driver (we might call this operator surveillance or response to alarm). Example: Front light beam unit failure.

b. Permanent Fault – It is a type of fault that persists even after attempts to correct or reset the system. that stays until repaired. Example: Thermal Runaway

c. Residual Fault – a "portion of a fault" that leads to system failure and is not covered by a safety mechanism.

d. Single Point Fault – a fault that directly leads to system failure and can be covered by a safety mechanism. Example: Fault in memory unit

e. Dual / Multiple Point Fault – a fault that does not directly lead to system failure but can lead to failure in combination with other faults. It can be covered by a safety mechanism. Example: Power monitor fault. It does not directly affect the system function but when power fluctuation fault occurs, it cannot be detected and hence failure occurs.

f. Safe Fault – a fault that will not lead to failure i.e. violation of Safety goal, in any operating condition Example: Fault in redundant memory

g. Detected Fault – fault that can be detected by a safety mechanism Example: Error in communication protected by CRC.

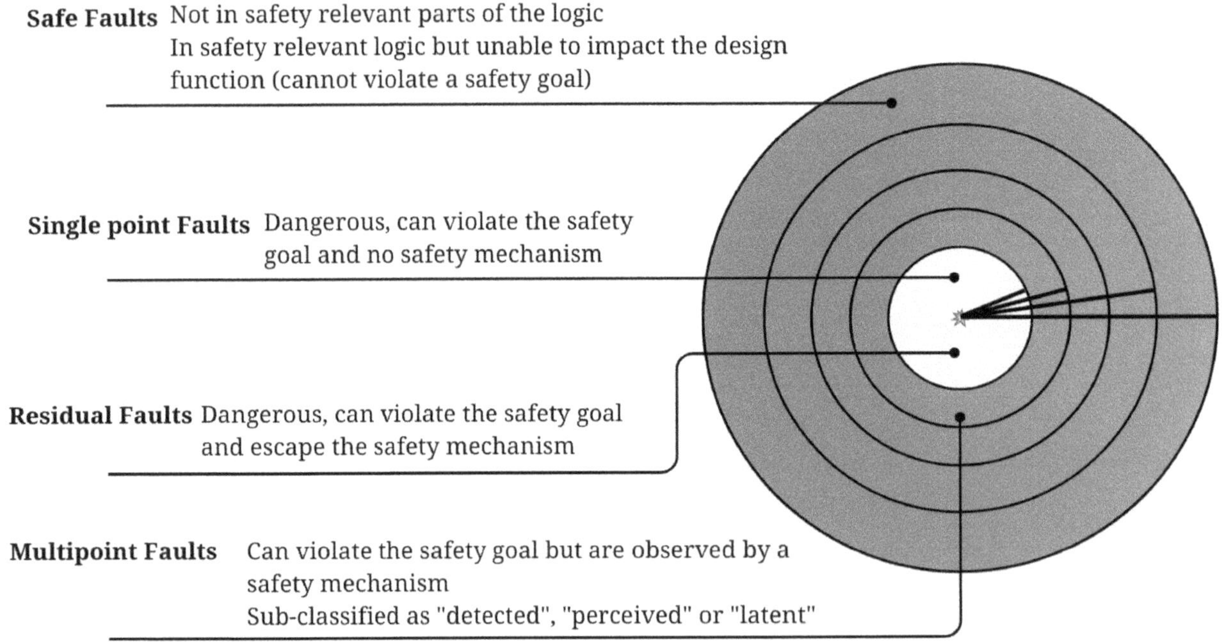

Failure - The interruption of the desired behavior of a product or products due to an adverse event. Termination may be permanent or temporary.

Failure Rate – The failure rate is divided by the lifetime of the hardware. Suppose the failure rate is constant and is usually defined by "λ".

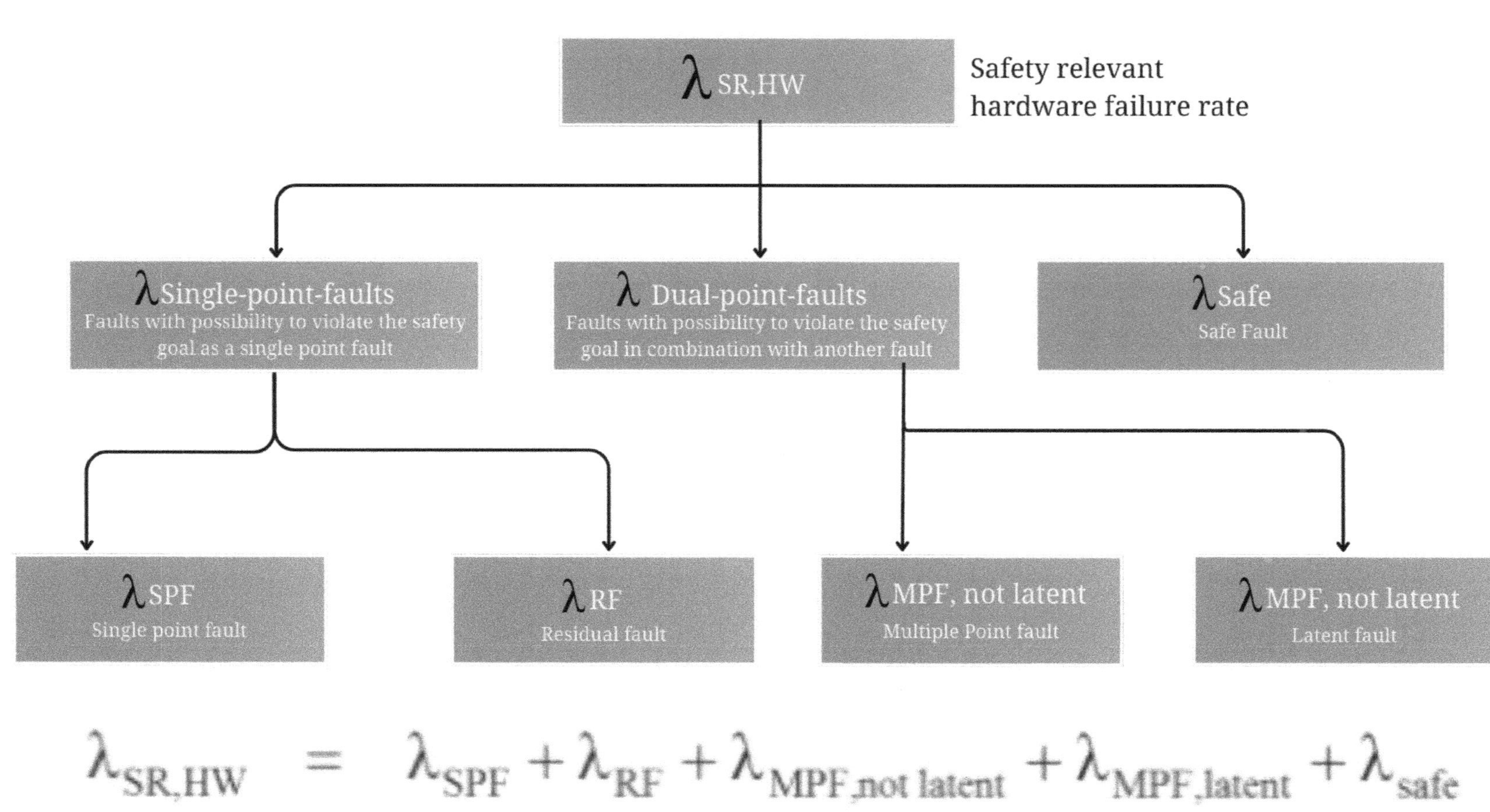

$$\lambda_{SR,HW} = \lambda_{SPF} + \lambda_{RF} + \lambda_{MPF,not\ latent} + \lambda_{MPF,latent} + \lambda_{safe}$$

The Safety Relevant Hardware failure rate of the overall system would be addition of failure rate of Single point faults, residual faults, multi-point faults and safe faults in the system. [5]

Systematic vs random faults/failures

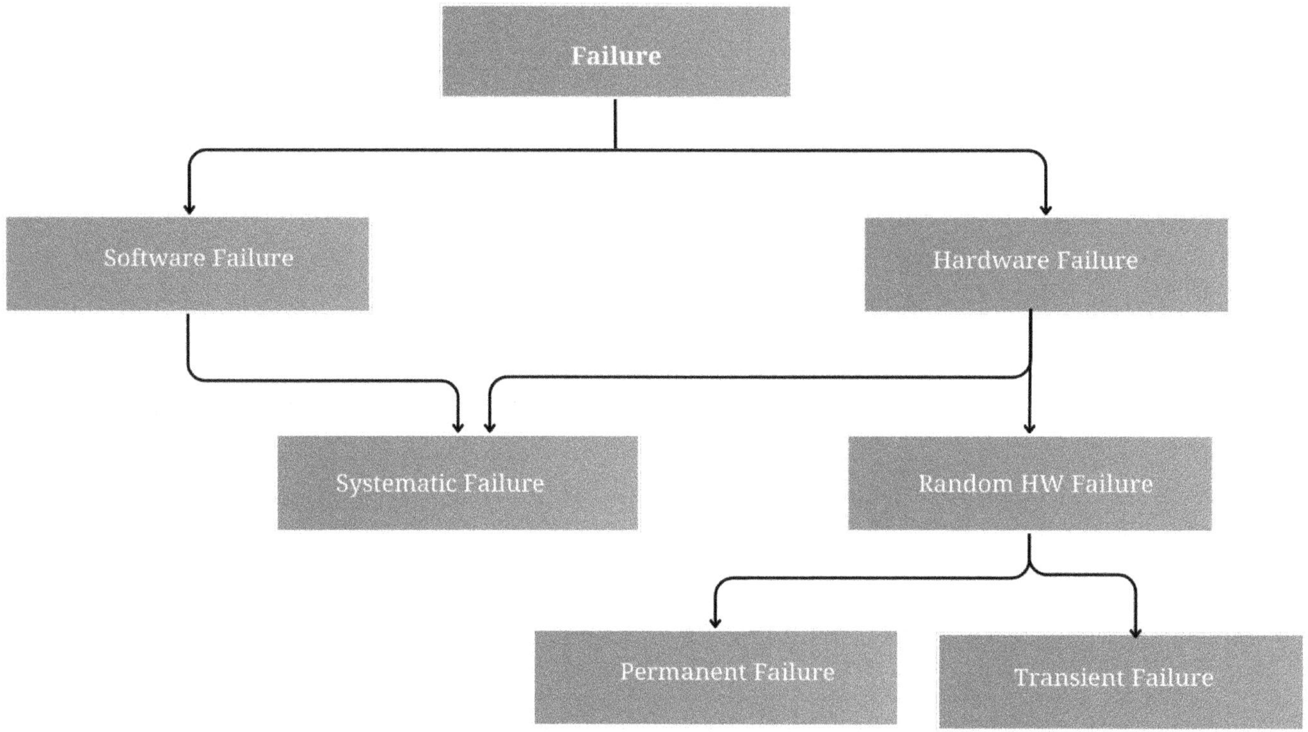

The failures can be highly classified into Hardware and Software failures. Systematic failures are faced during both Hardware and Software development. Random Hardware Failure is faced during the lifetime of the product by the Hardware components. Systematic failures are development errors which are deterministic, risks due to systematic failures can be prevented by developing Hardware and Software to correct design standards (guidelines) and perform safety analysis and system verification.

Risks due to Random hardware failures can be reduced by deploying the correct active safety mechanism (e.g. Built in tests) into the system (item).

Systematic Fault – fault whose failure is manifested in a deterministic way that can only be prevented by applying process or design measures

Types of Systematic Fault:

1. Transient fault:

a. Fault that occurs once and subsequently disappears

b. Transient faults can appear due to electromagnetic interference, which can lead to bit-flips.

c. Soft errors such as Single Event Upset (SEU) and Single Event Transient (SET) are transient faults.

d. Transient faults include the momentary loss of network connectivity to components and services, the temporary unavailability of a service, or timeouts that arise when a service is busy.

e. These faults are often self-correcting, and if the action is repeated after a suitable delay it is likely to succeed.

2. Safe fault:

a. Fault whose occurrence will not significantly increase the probability of violation of a safety goal

b. Both non-safety and safety-related elements can have safe faults.

c. Single-point faults, residual faults and dual-point faults do not constitute safe faults.

d. Unless shown relevant in the safety concept, multiple-point faults with higher order than 2 can be considered as safe faults

Types of System Failure

Fail-dangerous: This type of system failure is characterized by a dangerous or hazardous outcome when a failure occurs. It means that the failure of the system can result in severe consequences, such as injury or loss of life. Fail-dangerous systems typically lack appropriate safety measures or backup systems to mitigate the risks associated with a failure.

Fail-inconsistent: Fail-inconsistent refers to a system failure where the behavior of the system becomes inconsistent or unpredictable after the failure. The system may produce incorrect or unreliable outputs, making it difficult to rely on its functionality or make accurate decisions based on its outputs.

Fail-stop: A fail-stop failure refers to a situation where the system detects a failure and stops functioning entirely. In this case, the system halts its operation to prevent further damage or erroneous behavior. Fail-stop failures are often designed to ensure safety and prevent the system from continuing to operate in an erroneous or compromised state.

Fail-safe: Fail-safe systems are designed to minimize harm or damage in the event of a failure. When a failure occurs in a fail-safe system, it is designed to enter a safe state or mode of operation that ensures the system's integrity or the safety of users. Fail-safe mechanisms aim to mitigate the impact of failures and prevent catastrophic consequences.

Fail-operational: Fail-operational systems are designed to continue operating, albeit with degraded functionality or performance, even in the presence of failures. These systems have redundancy or backup mechanisms that allow them to maintain their essential operations despite the occurrence of a failure. Fail-operational systems are often used in critical applications where uninterrupted operation is vital.

Fail-silent: Fail-silent refers to a type of failure where the system continues to operate but fails to produce any indication or alert of the failure. In other words, the system fails without providing any observable signals or notifications of the failure. This can make it challenging to detect the failure and take appropriate corrective actions.

Fail-indicate: Fail-indicate systems are designed to provide clear and unambiguous indications or alerts when a failure occurs. These indications can be visual, auditory, or through other means to notify users or operators of the system about the failure. Fail-indicate mechanisms help in promptly identifying and addressing failures, facilitating timely troubleshooting or repair.

These types of system failures are often considered during the design and analysis phase to ensure that appropriate measures are in place to handle failures and mitigate their impact.

Notice that the first two fail scenarios are undesired; after implementing safety mechanisms and TSRs they will be converted into any combination of the last 5 fail-safe types. In other words, you can specify your safe state to be operational and indicate or stop and be silent.

Random Fault – May occur unpredictably during a lifetime of a Hardware element. Hardware element failure rate is usually expressed with the below bathtub curve. Observed failure rate of any hardware element over its lifetime period traces a bathtub shape.

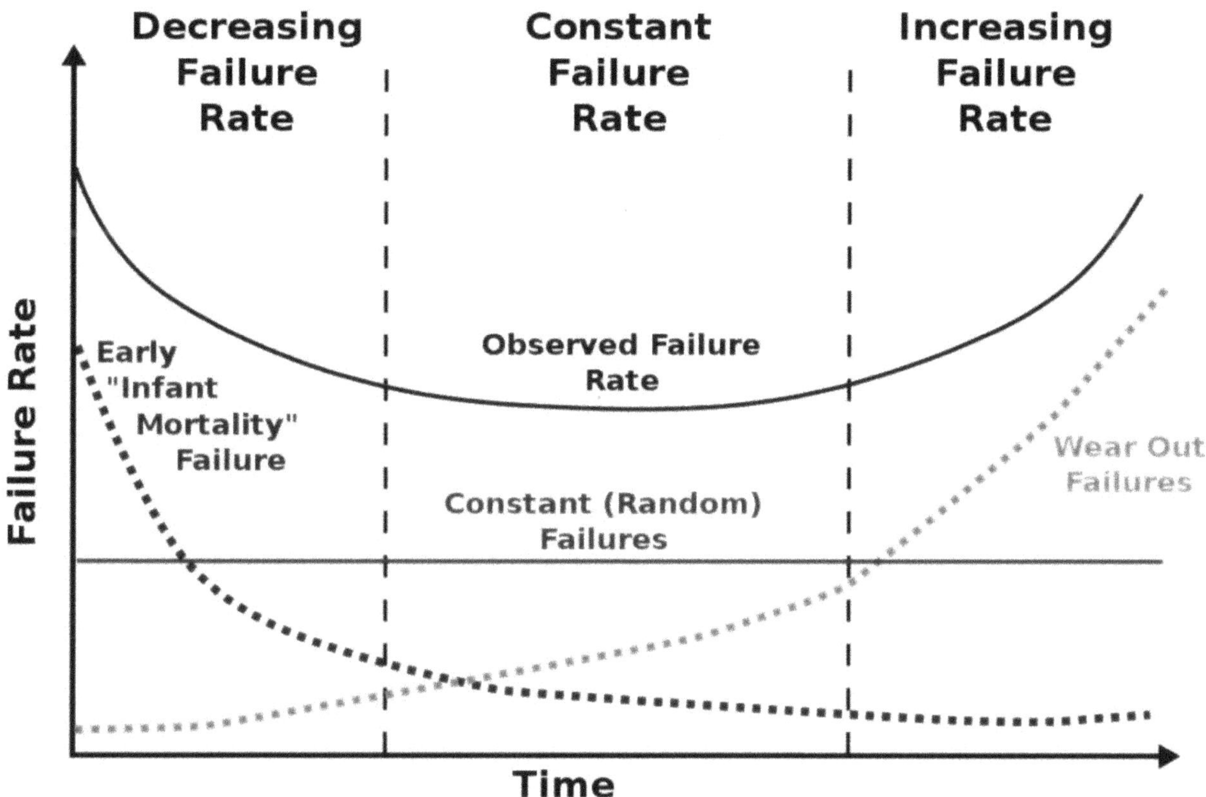

The random hardware failure over time consists of three phases:

1. The first phase - In this phase, failure due to manufacturing and processing defects happen. Hence called a period with high "Infant Mortality" rate. Since the product is new, it will be failing during the fabrication process and during component screening.

2. The Second Phase - This is the useful life period. Useful life period is during the period where constant failure rate is expected and is linked to intrinsic failures. The failure rate is assumed to be constant during this time either for an unlimited period of operation, or for limited periods of operation. This is the phase where the product is predicted to be under function at the end customer. ISO26262 analysis and requirements are to lower the failure rate at this phase.

3. The Third Phase - This phase will again have increased probability of failures due to wear-out of components. This will be tracked by the End-of-Life period of the components used.

Types of Random Hardware Failure

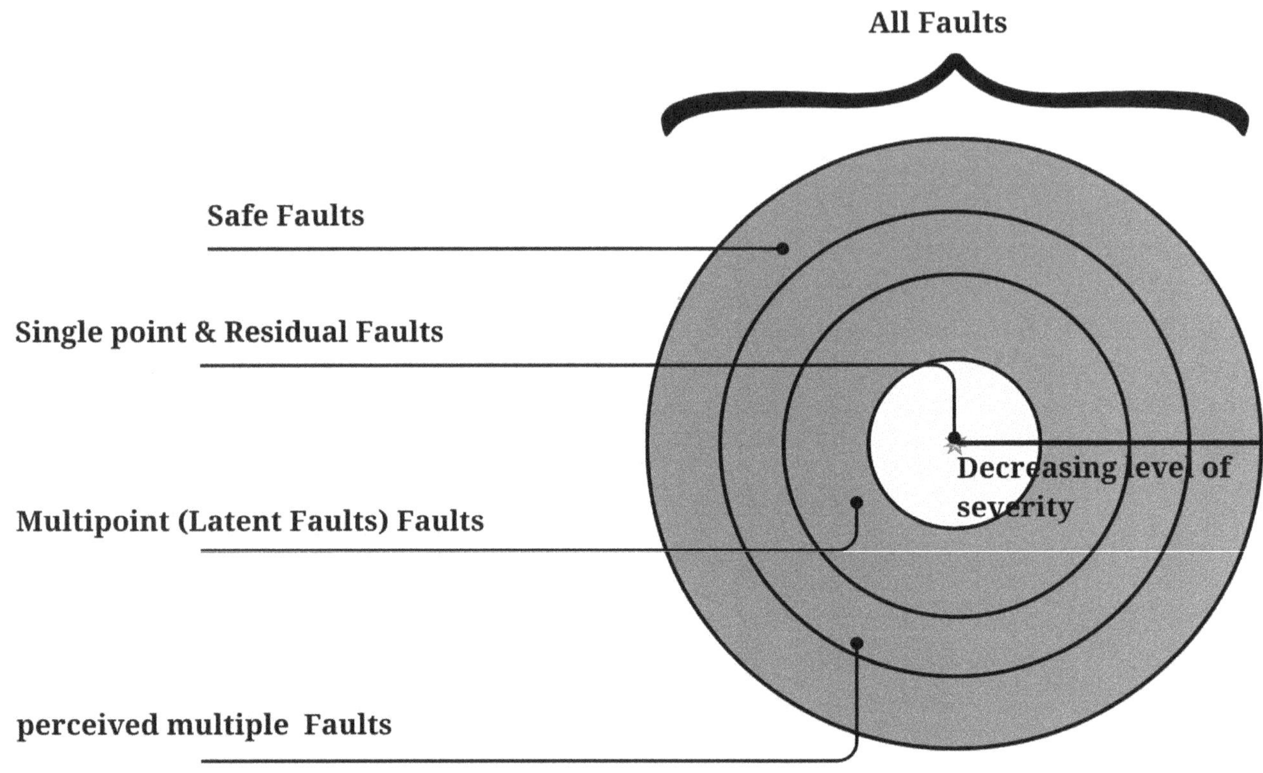

Let us revisit the definition of the different types of faults, here a little more in detail:

1. Single Point Fault:

    a.   Hardware fault in an element that leads directly to the violation of a safety goal and no fault in that element is covered by any safety mechanism.

    b.   If at least one safety mechanism is defined for the hardware element, then no faults of the considered hardware element are single point faults. The remaining faults that are not covered by the present safety mechanism will be the residual faults.

Example: Open of a resistor which can lead to a violation of a safety goal, which is not monitored by any Safety mechanism. If monitoring by microcontroller is implemented, this will be a Safety mechanism and the remaining percentage of fault which is not addressed by the monitoring of microcontroller will be the residual fault.

2. Dual-point fault/Latent point fault:

   a. Individual fault that in combination with another independent fault leads to a dual-point failure

   b. Latent fault is a dual-point fault whose presence is not detected by a safety mechanism nor perceived by the driver within the multiple-point fault detection time interval

   c. Example: Watchdog failure is a dual point fault. This will not affect the safety goal or working of a microcontroller. The system can fully function properly without the watchdog. But, say, we have a RAM error in the microcontroller and the program counter memory is stuck, the watchdog will fail to reset the microcontroller and hence the whole system will be stuck, which will lead to violation of safety goal in many cases. If we get RAM error alone, the system is going to have a functioning watchdog to reset and recover the system. Watchdog failure along with the program counter memory failure leads to violation of safety goal, requiring two points to fail, hence known as a dual-point fault.

3. Multiple-point fault: Individual faults that, in combination with other independent faults, if undetected and not perceived, could lead to a multiple-point failure. Multiple point fault can be in a 'perceived' state or a 'detected' state.

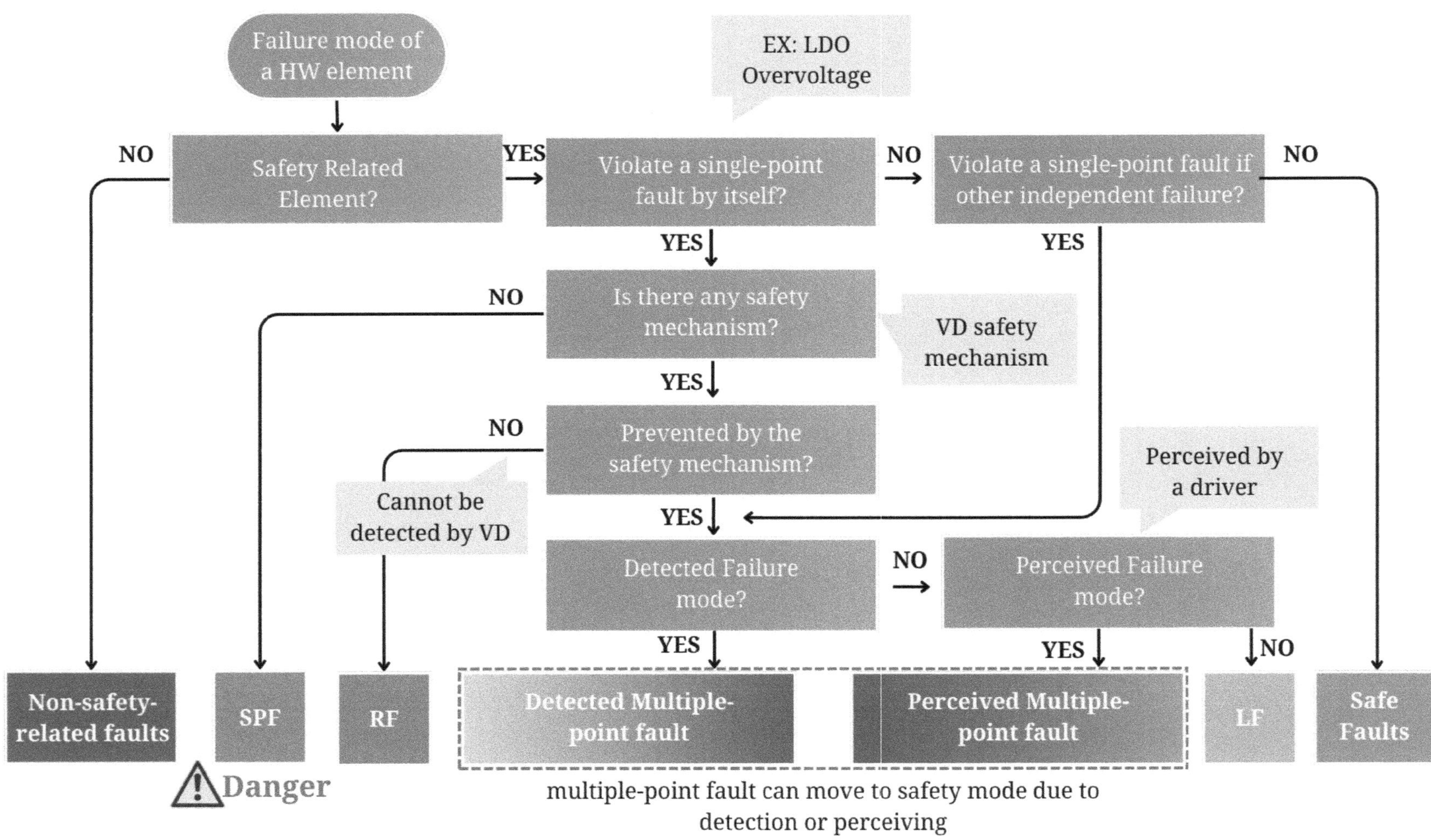

Failure mode of a HW element
Safety Related Element?
NO
YES
Violate a single-point fault by itself?
EX: LDO Overvoltage
NO
Violate a single-point fault if other independent failure?
YES
NO
YES
Is there any safety mechanism?
VD safety mechanism
NO
YES
Prevented by the safety mechanism?
NO
Cannot be detected by VD
YES
Detected Failure mode?
NO
Perceived Failure mode?
Perceived by a driver
YES
YES
NO
Non-safety-related faults
SPF
RF
Danger
Detected Multiple-point fault
Perceived Multiple-point fault
LF
Safe Faults
multiple-point fault can move to safety mode due to detection or perceiving

4. Residual fault:

a. Portion of a random hardware fault that by itself leads to the violation of a safety goal, occurring in a hardware element, where that portion of the random hardware fault is not controlled by a safety mechanism

b. This presumes that the hardware element has safety mechanism coverage for only a portion of its faults.

c. If a set of faults which is safety-relevant and not safe has a subset with 60 % coverage, then the remaining 40 % of the set of faults are residual faults.

Example: A communication mechanism could have the following failure modes:

i. Loss of communication peer

ii. Message corruption

iii. Message unacceptable delay- Message loss

iv. Unintended message repetition

v. Incorrect sequencing of messages

vi. Message insertion

vii. Message masquerading

viii. Message incorrect addressing

Let us assume that we introduce a SW safety mechanism that addresses all of the failure modes except Message Masquerading, which remains unaddressed, it is considered as a Residual Fault

Dependent vs independent failures

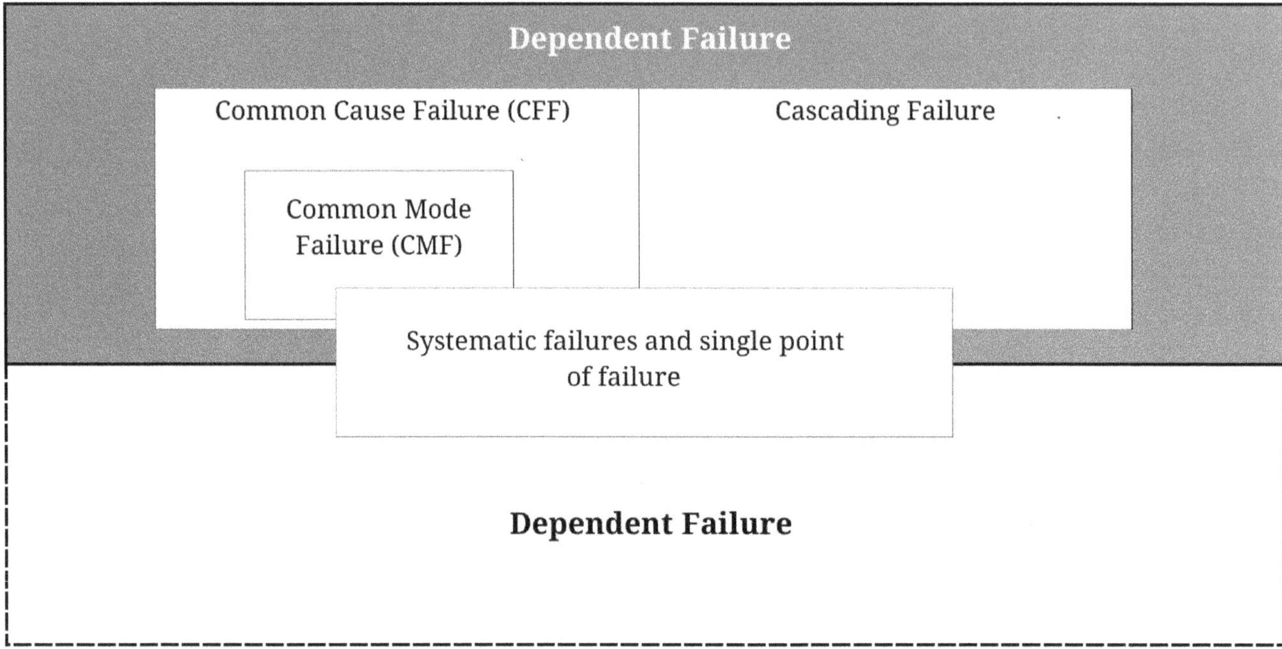

Independent failure – failures whose probability of simultaneous or successive occurrence can be expressed as the simple product of their unconditional probabilities. Example: Let us say we have two temperature sensor in the system. Temperature sensor A internally fails, this will not affect Temperature sensor B. Also, the failure can be detected by the microcontroller unit and necessary actions can be taken. It does not lead to failure of any other element in the system.

Dependent Failure – failures that are not statistically independent, i.e. The occurrence of simultaneous or sequential events cannot be shown as a simple product of the unsustainable nature of each person's failure. Dependent failures can exhibit themselves simultaneously, or within a sufficiently short time interval, to have the effect of simultaneous failures. Dependent failures include common cause failures and cascading failures. Whether a given failure is a cascading failure or a common cause failure may depend on the hierarchical structure of the elements and on the temporal behaviors of the elements.

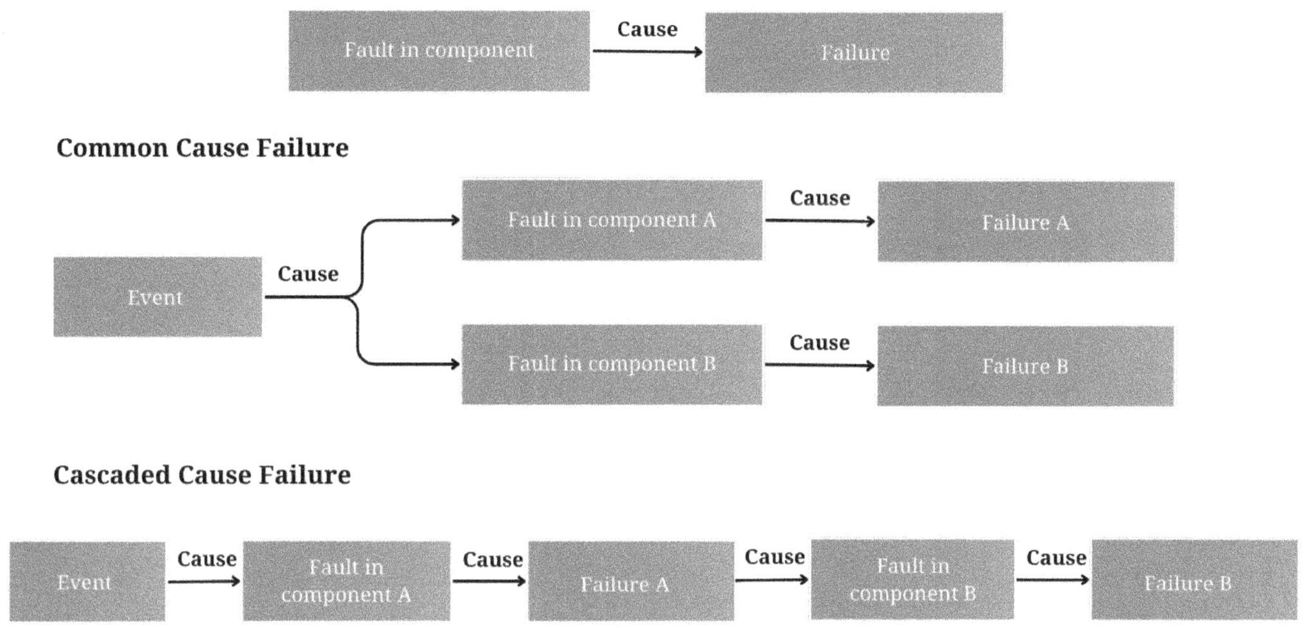

Cascading Failure – Failure of an element of an item resulting from a root cause inside or outside of the element and then causing a failure of another element or elements of the same or different item. Cascading failures are dependent failures that could be one of the possible root causes of a common cause failure.

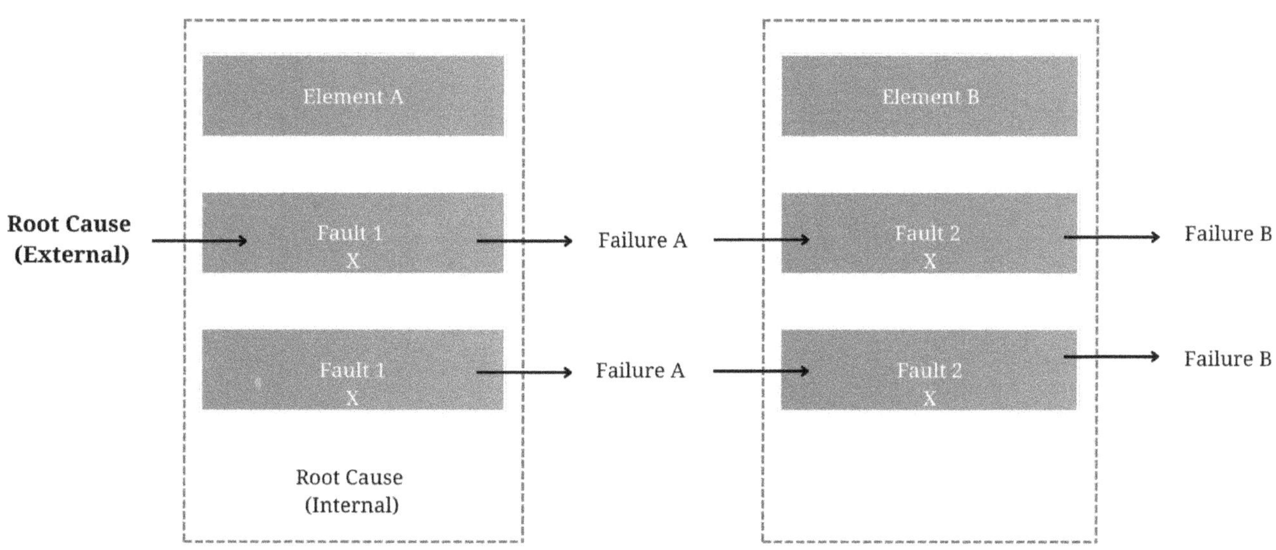

Common Cause Failure – Failure of two or more elements of an item resulting directly from a single specific event or root cause which is either internal or external to all of these elements

Common cause failures are dependent failures that are not cascading failures.

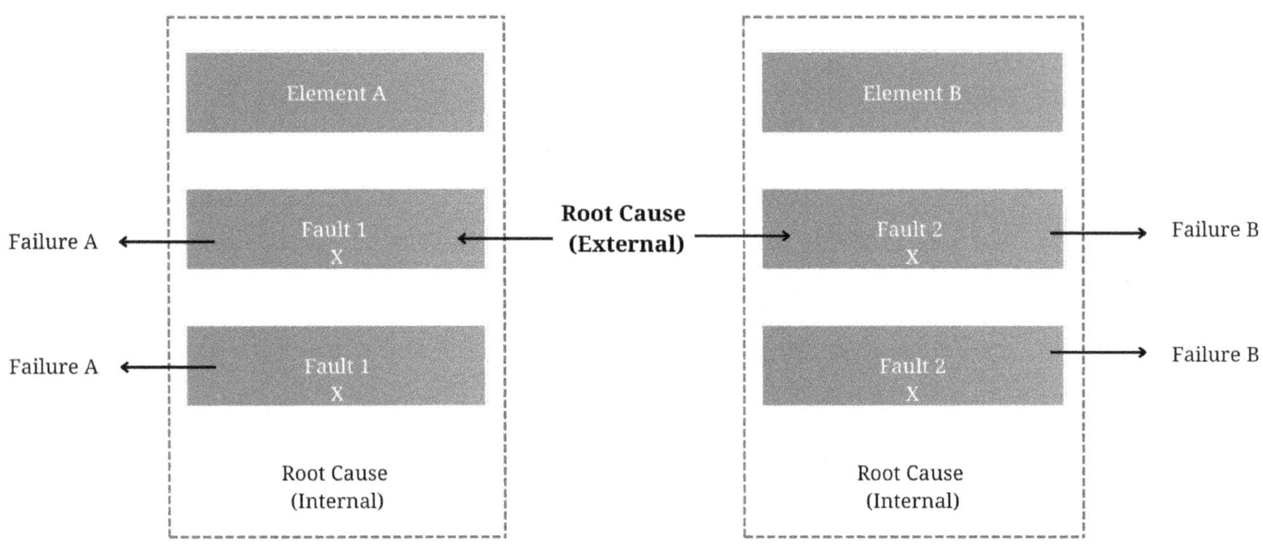

Here, the power supply failure would be a common-cause failure, as we can see if the regulator fails, then both MCU fails. A Cascading failure can be, for example, a memory failure of MC1, this will not affect MC2 or any other component simultaneously, but the output of MC1 will fail, hence the components functioning based on MC1 output will fail, hence the overall system output will fail. As the name provides, the failure cascades.

Terms for ISO26262 metrics

FIT – Failure In Time. In ISO26262 a FIT is 10E-9 failures per operational hour.

The reliability of a Hardware part is the probability of performing a required mission without failure, under stated conditions for a stated mission time. This reliability is expressed in FIT in context of this standard. The Failures In Time (FIT) rate of a device is the number of failures that can be expected in one billion (10E+9) device-hours of operation

This is calculated with expressions defined by the IEC62380 standard for Reliability calculation. There are also other standards such as SN29500 - Siemens, MIL-HDBK-217F that can be used. These standards provide the probability of failure of a component in the Useful life period, based on the characteristics of the component such as power dissipation, temperature resistance etc. of the component.

The FIT value of a Hardware part is determined based on the parameters of the component (such as the power dissipation, package, thermal resistance) and the mission profile of the Element (Exposure to temperature ranges, Percentage of hours the element will be exposed to a particular temperature, whether the component is placed in passenger compartment or the engine compartment)

Mission Profile – Reliability calculation of equipment has to be done according to its field use conditions. They are defined by the mission profile which is nothing but the temperature profile in which the component will be used in the field. It defines parameters such as exposure to temperature ranges outside average temperature, Percentage of hours the element will be exposed to a particular temperature, whether the component is placed in passenger compartment or the engine compartment, temperature swings that will happen during on-off cycle, day-night conditions, etc.

Diagnostic Coverage – percentage of the failure rate of a hardware element, that is detected or controlled by the implemented safety mechanism. Diagnostic coverage can be assessed with regard to residual faults that might occur in a hardware element after implementing the Safety Mechanism.

Single-point Fault Metric – measures the robustness of the design to single-point and residual faults. This metric shows how much the single-point faults of the design are covered by Safety mechanisms and its diagnostic coverage. The requirement as per ISO26262 on how much the single-point faults shall be covered for the design to meet respective ASIL is given by below table:

Table 4 — Possible source for the derivation of the target "single-point fault metric" value

	ASIL B	ASIL C	ASIL D
Single-point fault metric	≥90 %	≥97 %	≥99 %

The single-point fault metric is defined as 1 minus "the sum of the single-point faults and the residual faults of the safety relevant hardware parts" divided by "the sum of failure rate of all safety relevant hardware parts."

$$SPFM = 1 - \frac{\sum_{SR,HW} \left(\lambda_{SPF} + \lambda_{RF} \right)}{\sum_{SR,HW} \lambda}$$

Where, SPFM is the Single-point fault metric

SR,HW is the Safety Relevant Hardware parts

λSPF is the total failure rate of Single Point Faults

λRF is the total failure rate of Residual Faults

λ is the total failure rate of all Hardware Components

Latent Fault Metric –

The latent fault metric is defined as 1 minus "the sum of the latent multiple-point faults of the safety relevant hardware parts" divided by "the total failure rate of all safety relevant hardware parts excluding total failure rate of single-point faults and residual faults" i.e. the following ratio:

$$LFM \;=\; 1 - \frac{\displaystyle\sum_{SR,HW} \left(\lambda_{MPF,L}\right)}{\displaystyle\sum_{SR,HW} \left(\lambda - \lambda_{SPF} - \lambda_{RF}\right)}$$

Where, LFM is the Latent fault metric

SR,HW is the Safety Relevant Hardware parts

λMPF,L is the total failure rate of Multi-Point Faults that have Latent effect(i.e. total failure rate of all Dual-point faults)

λSPF is the total failure rate of Single Point Faults

λRF is the total failure rate of Residual Faults

λ is the total failure rate of all Hardware Components

The requirement as per ISO26262 on how much the latent faults shall be covered for the design to meet respective ASIL is given by below table:

Table 5 — Possible source for the derivation of the target "latent-fault metric" value

	ASIL B	ASIL C	ASIL D
Latent-fault metric	≥60 %	≥80 %	≥90 %

Probabilistic Metric for random Hardware Failures (PMHF) - it is a metric that expresses the total probability of the random hardware failure rate.

The maximum failure probability of element that can be tolerated for each ASIL is given by the ISO26262 as below:

Table 6 — Possible source for the derivation of the random hardware failure target values

ASIL	Random hardware failure target values
D	$<10^{-8}\ h^{-1}$
C	$<10^{-7}\ h^{-1}$
B	$<10^{-7}\ h^{-1}$

NOTE   The quantitative target values described in this table can be tailored as specified in 4.2 to fit specific uses of the item (e.g. if the item is able to violate the safety goal for durations longer than the typical use of a passenger car).

Note: The Hardware metrics SPFM, LFM and PMHF are not provided for ASIL A as the ISO26262 demands the hardware metric analysis or the quantitative analysis only for ASIL B and higher ASILs.

Important Abbreviations in ISO26262

The document Part 1 of ISO26262 gives an ample list of abbreviations. Some of the new abbreviations used in the standard listed below:

Abbreviation	Description
ASIL	Automotive Safety Integrity Level
BFR	Base Failure Rate
BIST	Built-In Self-Test
CCF	Common Cause Failure
COTS	Commercial Off The Shelf
DC	Diagnostic Coverage
DFA	Dependent Failure Analysis
DFI	Dependent Failure Initiator
DIA	Development Interface Agreement

Abbreviation	Description
EEC	Evaluation of Each Cause of safety goal violation
EOTI	Emergency Operation Time Interval
EOTTI	Emergency Operation Tolerance Time Interval
ESC	Electronic Stability Control
ETA	Event Tree Analysis
EVR	Embedded Voltage Regulator
FDTI	Fault Detection Time Interval
FHTI	Fault Handling Time Interval
FIT	Failures In Time
FMC	Failure Mode Coverage
FMEA	Failure Mode and Effects Analysis
FRTI	Fault Reaction Time Interval
FTA	Fault Tree Analysis
FTTI	Fault Tolerant Time Interval
HARA	Hazard Analysis and Risk Assessment
HAZOP	HAZard and OPerability analysis
HSI	Hardware-Software Interface
HS/LS	High Side / Low Side
LFM	Latent-Fault Metric
MBD	Model Based Development
MC/DC	Modified Condition/Decision Coverage
OV	Over Voltage
PMHF	Probabilistic Metric for random Hardware Failures

Abbreviation	Description
PPAP	Production Part Approval Process
PTO	Power Take-Off
QM	Quality Management
RF	Residual Fault
RFQ	Request For Quotation
SEB	Single Event Burnout
SEE	Single Event Effect
SEGR	Single Event Gate Rupture
SEooC	Safety Element out of Context
SET	Single Event Transient
SEU	Single Event Upset
SG	Safety Goal
SOP	Start Of Production
SPFM	Single-Point Fault Metric
T&B	Trucks, Buses, trailers and semi-trailers
TCL	Tool Confidence Level
TD	Tool error Detection
TI	Tool Impact
UV	Under Voltage

Chapter 3

ISO26262 Part 2 – Safety Management

Topics to be covered

1. What is described by Part 2 of ISO26262?

2. Functional safety management in a company as per ISO26262

3. Overall safety management of a project as per ISO26262

4. Project – dependent and project independent safety management

5. Safety management after release for production

6. Work products required by Part2 ISO26262

7. Functional Safety Lifecycle

8. Interpretation of Tables and requirement applicability

Purpose of Part 2 of ISO26262

The purpose of Part 2 of ISO26262[1] is to instruct how a project should be managed to make it compliant with ISO26262 Functional Safety Standard. The clause provides instructions on how to Achieve compliance with the standard and how to create the evidence of Safety implementation in each phase of the project - The project specific implementation has to be planned in functional safety management phase in the work product, Safety Plan.

How to interpret tables used in the ISO26262 standard and How to interpret the upcoming clauses (Part 3 to 12) of ISO26262 to know their applicability depending on ASIL level of the product and its function.

Functional Safety Management as per ISO26262

ISO26262 has defined a set of rules for a company if it needs to develop and establish a functional safety team which primarily functions on certifying the projects to meet a defined ASIL level to comply with ISO26262 standard. We will see the requirements in detail below:

Overall safety management requires that the company must have defined and applied procedures for electronics development.

This includes procedures

- to define a company-specific lifecycle,

- to define which tools to use,

- on how to organize configuration management and

- on which kinds of safety analyses to perform for any project respective to ASIL

Management of functional safety needs to be addressed at

- on an organizational level, (overall safety management project independent)

- on a project level (project dependent safety management) and

- for the time after release for production.

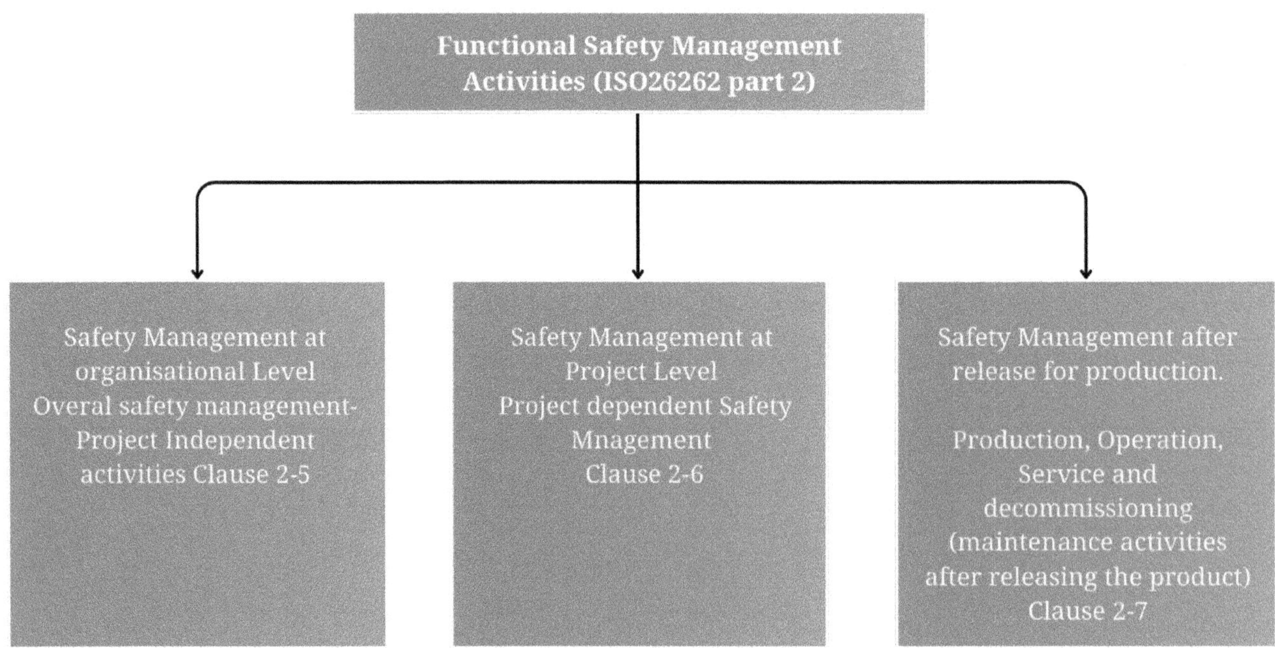

Overall Safety Management

Functional Safety Management at organizational level.

- Must establish a good management system. This is also supported by the Automotive SPICE standard. Automotive SPICE supports compliance with ISO 26262.

- Have a process in Functional Safety, to ensure that safety anomalies are identified, communicated, and resolved.

- 'culture of safety' is the term defined by people– which every company should have Safety culture and processes at the organization

An organization that develops electronic products for cars must have an organized safety culture and defined processes for functional safety.

Competence management

Additionally, ISO 26262 requires that you are qualified for the work assigned to you. This means that there must be active competence management in your organization. There must be certified training that shall be organized for the safety engineers and safety managers to be trained for their roles and responsibilities.

Requirements on Safety Culture - Roles and responsibilities

- Organization Structure in Functional Safety Planning

- Achieving functional safety in automotive development needs all the stakeholders to work towards this common goal.

- Interactions between team members should be defined in the safety planning activity sheet. Role definition, who interacts with whom, who advances to whom, etc. matters should be clearly stated in a section of the safety planning activity sheet.

A possible FSM (Functional Safety Management) structure is given below:

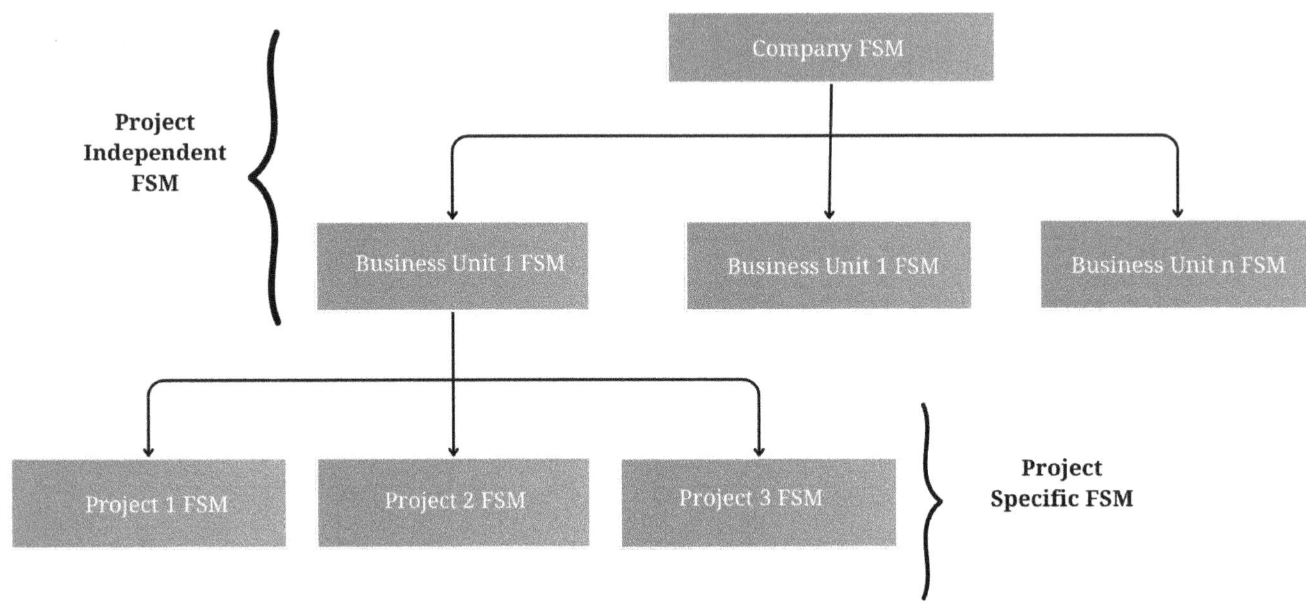

Project Roles & Responsibilities

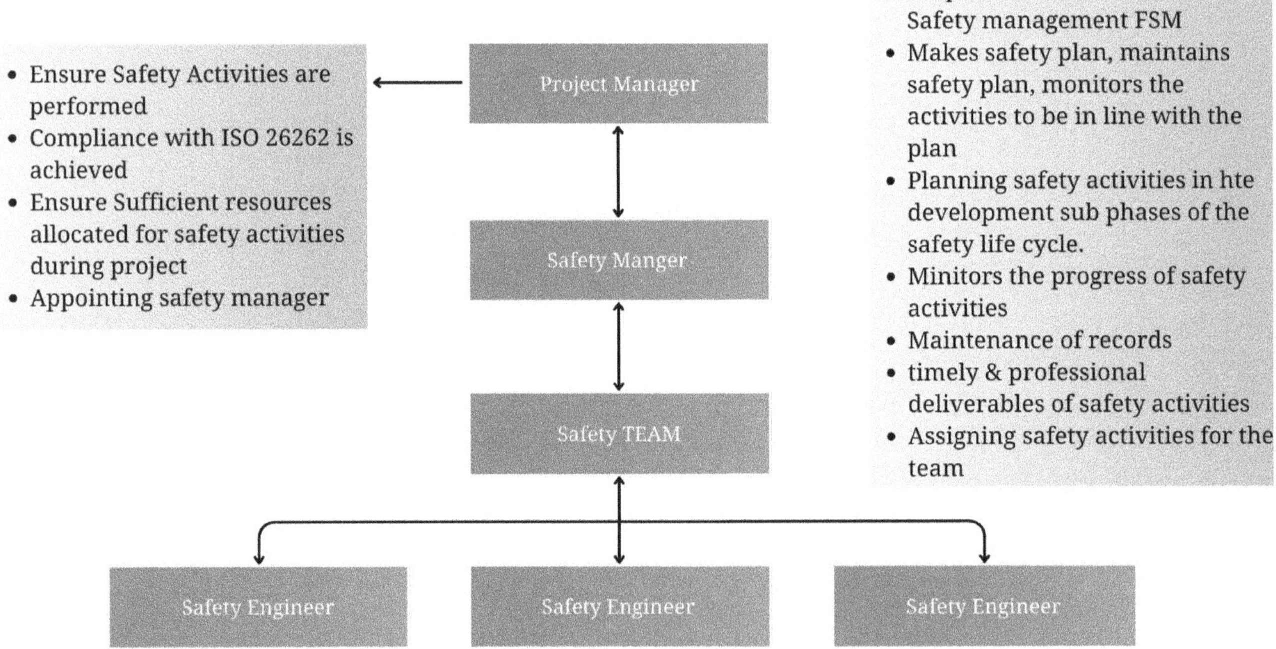

Project Manager

Part 2 Section 6 of ISO 26262[1] document recommends that a project manager should be appointed at the initiation of a functional safety project, who ensures the project is developed in compliance with ISO26262 by ensuring it meets the defined safety goals and all safety activities are performed as per the ASIL definitions.

Functional Safety Manager

A functional safety manager also needs to be appointed who will be responsible for following activities related to safety planning and coordination:

- Delegation of task based on skill set, competence, and qualification

- Maintenance of the safety plan throughout the safety lifecycle

- Monitoring the progress of all the automotive functional safety activities

- Timely delivery of work-products for each phase of the project

- Maintenance of delivered work-products for future references and audit/assessment purposes

Safety Engineer

A safety engineer will be responsible for carrying out delegated tasks. Safety engineers are human resources provided to the safety manager to execute the safety activities. It is the responsibility of the safety manager to ensure the safety engineer is trained and is executing the delegated safety activities correctly and in a timely manner.

Project Resources at Work

In addition to the human resources, software tools, databases, templates etc. are also required to achieve functional safety goals. It is organization's duty to provide all such resources. And Safety manager has to ensure that human resources has access to them and to request for access in case of any lack. It is the role of project manager to arrange the resources in case of any lack of accessibility.

Functional safety management at the project level - project dependent safety management

In Safety management, there are two types of activities

- Project-dependent

- Project-independent

Project independent Safety management

Project independent safety management activities will not change with respect to the project description. These will cover the below mentioned activities.

- Implement Quality Management System (IATF 16949 in conjunction with ISO 9001)

- Safety Culture

- Management of safety anomalies regarding functional safety

- Competence management

- Project-independent tailoring of the safety lifecycle

- Establish Company wide safety lifecycle including the supporting processes

- Define Process for continuous Improvement of functional safety

You can note that almost all the parts are covered by the previous topic. So the project-independent safety management would be the organizational topics which has to be ensured to build an eligible team for qualifying projects to meet ISO26262 standard.

Project dependent Safety Management

Project dependent safety management would be where we will be taking in the project and work with the eligible team to qualify the given target project to meet ISO26262 certification.

Below are the activities:

1. Roles and responsibilities in safety management

2. Impact analysis at the item level

3. Reuse of an existing element

4. Tailoring of the safety activities

5. Planning and coordination of the safety activities

6. Progression of the safety lifecycle

7. Safety Case

8. Confirmation measures

9. Confirmation review

    a. Functional safety audit

    b. Functional safety assessment

10. Release for production

Roles and responsibilities in safety management

As we discussed in the previous topic, the project manager, Safety manager and safety engineers have to be assigned for this particular project and the necessary resources have to be allocated.

Impact analysis at the item level

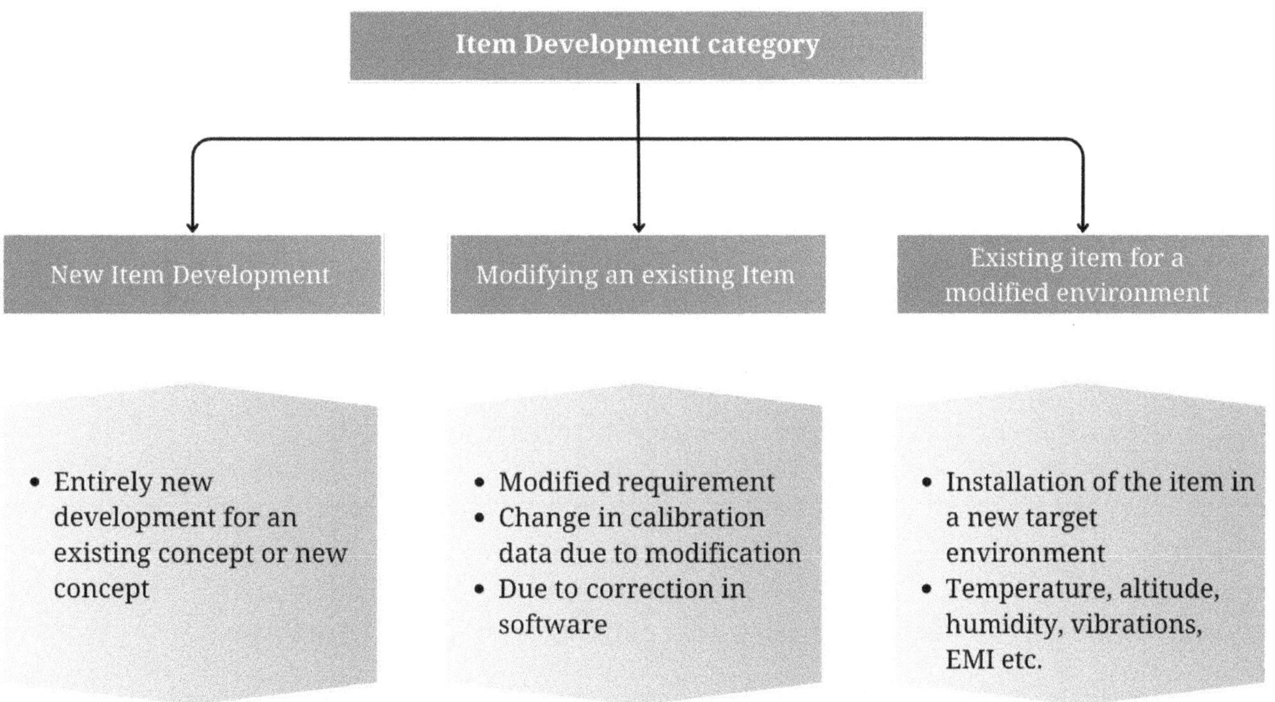

An impact analysis conveys what needs to be done in the process of product development. Because only in the rarest of cases, a company starts a product development from scratch. There is always a previous version of the product, as shown in the above figure. The result of the impact analysis is used

- to tailor the product lifecycle and

- to plan the important safety activities.

Reuse of an existing element:

An existing element can be reused:

a. based on an evaluation of hardware elements (ISO 26262-8:2018, Clause 13),

b. based on a qualification of software components (ISO 26262-8:2018, Clause 12),

c. based on a proven in use argument (ISO 26262-8:2018, Clause 14), or

d. as a Safety Element out of Context (ISO 26262-10, Clause 9).

Tailoring of the safety activities

Safety activities related to the development of a particular item can be changed, for example removed, or implemented in a different way than specified in the implementation of the ISO 26262 lifecycle.

- A justification should be given as to why tailoring is necessary and sufficient for safety.

- Logic considers compliance ASIL requirements.

- Interrupt considerations are included in the safety plan and reviewed during the actual audit or security assessment of the safety plan.

- This requirement applies to project-specific application adaptation.

- With regard to tailoring of the safety lifecycle for application across item developments within an organization, only Project independent tailoring applies

- This requirement applies to item developments for Trucks and Buses

Planning and coordination of the safety activities - Safety Plan

This Safety Plan is a work-product defined by Part2 of ISO26262. This document shall be drafted, reviewed and approved. This has to be assessed and delivered to the customer. This document provides the list of tasks required as part of Safety Life Cycle according to ISO 26262 process and many more details. The Safety plan will be detailed further in the upcoming chapter of this document.

Progression of the safety lifecycle

- In the case of a lack of information from the pertinent preceding sub-phases, a subsequent sub-phase shall only start if the lack of information does not cause an unreasonable risk regarding functional safety

- For cases where the lack of information can jeopardize the project, the issue is escalated

- The work products required by the safety plan shall be subject to configuration management, change management and documentation no later than the time of entering the phase "product development at the system level"

Safety Case

This document would be the evidence of functional safety implementation and that the project development has followed ISO26262 project and the project has met a particular ASIL rating. The Safety case will be detailed further in the upcoming chapter of this document.

Confirmation measures

Functional safety also means restricting a project from misinterpreting ISO 26262 and cheating. This is to be confirmed through confirmation measures carried out by independent parties.

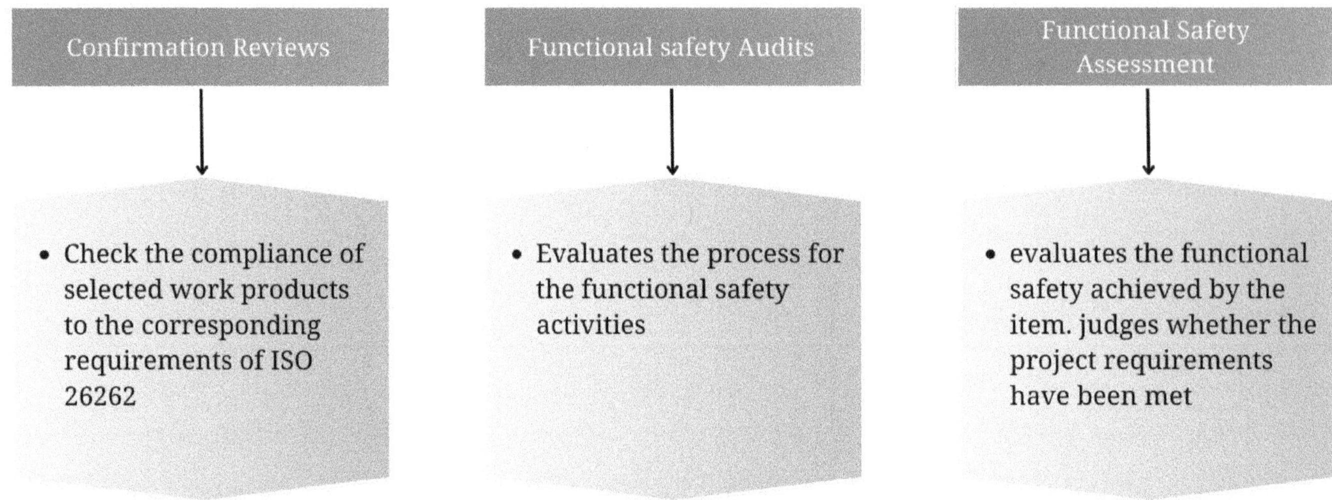

There are three types of confirmation measures.

- Verify reviews of key business products, such as safety plans or safety details. They need approval from an independent source.

- For a functional safety audit, it has to be checked whether the project actually implements the necessary and defined procedures. The audit therefore relates to compliance with the processes.

- An independent safety assessment is required for production. This is called a security assessment. This is required for high security products such as cruise control. Functional safety assessment is mandatory for ASIL C and D activities.

Development results released for production only.

- If a safety case occurs,

- if functional safety is independently recognized, and

- if development results are based on control settings upgrade

Independence of the reviewer

There are four levels of independence. The independence of the reviewer that is required for the confirmation measure of each safety activity is provided by ISO26262 as per the ASIL requirement of the project.

— I0: the confirmation measure should be performed; however, if the confirmation measure is performed, it shall be performed by a different person in relation to the person(s) responsible for the creation of the considered work product(s);

— I1: the confirmation measure shall be performed, by a different person in relation to the person(s) responsible for the creation of the considered work product(s);

— I2: the confirmation measure shall be performed, by a person who is independent from the team that is responsible for the creation of the considered work product(s), i.e. by a person not reporting to the same direct superior;

— I3: the confirmation measure shall be performed by a person who is independent, regarding management, resources and release authority, from the department responsible for the creation of the considered work product(s).

Confirmation review

- A confirmation review may be based on performing a judgment of whether the corresponding

- objectives of the ISO 26262 series of standards are achieved

- A person responsible to perform the confirmation review shall be appointed, for each confirmation review that is required by the safety plan. This person shall provide a report that contains a judgment of the achieved contribution to functional safety by the work product

- To increase confidence in the achievement of the review objectives, the reviewer checks the correctness, completeness, consistency, adequacy and contents of the work product against the corresponding requirements of the ISO 26262 series of standards

Functional safety audit

A functional safety audit may be based on a judgment of whether the process related objectives of the ISO 26262 series of standards are achieved

The person responsible to carry out a functional safety audit shall provide a report that contains a judgment of the implementation of the processes required for functional safety, based on:

a.   an evaluation of the implementation of the processes against the definitions of the activities referenced or specified in the safety plan;

b.   an evaluation of the safety plan products against the organization-specific rules and processes

c.   an evaluation of the arguments, if provided, as to why the process related objectives of the ISO 26262 series of standards are achieved;

Functional safety assessment

A functional safety assessment may be based on a judgment of whether the objectives of the ISO 26262 series of standards are achieved.

The achievement of an objective of the ISO 26262 series of standards is concluded considering the interelated requirements of these standards, the state-of-the-art regarding technical solutions and the pertinent engineering domain knowledge, at the time of the development.

d.   should be planned at the latest at the beginning of the product development at the system level;

e.   should be progressively performed during the product development; and

f.   shall be finalized before the release for production.

Release for production

A baseline for software and a baseline for hardware shall be available during release for production In addition, the same shall be under configuration management.

The release for production will be decided and approved by delivering the final safety case document and a valid pass result in the functional safety assessment and all other confirmation measures.

After the release of the documents, the product will be moved to the production phase where the production will be manufactured in multiples as per the customer requirement. The production process and after-market will be handled by the next clause.

Functional safety Management in Production and Aftermarket

If your development result is safe using safety management, you must ensure that the product is still built correctly and remains safe for the lifetime of the vehicles. ISO 26262 requires companies to register the names of their employees for production and field maintenance and to plan and initiate appropriate Safety measures. Appropriate procedures should be in place for this purpose.

Some control steps in production, for example reading flashed software to make sure there is no minor error in the controller. The on-site inspection process should ensure that the return to the field is audited for violations of safety objectives. Finally, safety operations require updating the software or hardware in the vehicle using on-site maintenance and spare parts replacement and replacement as needed.

To ensure the safety of the work not only during production and manufacture of the car, but throughout the life of the car, it is necessary to be well prepared, including the process of tracking location and change. Evidence/documentation should be created for the planning process for production, operation, service and transportation.

Work products defined by Part 2

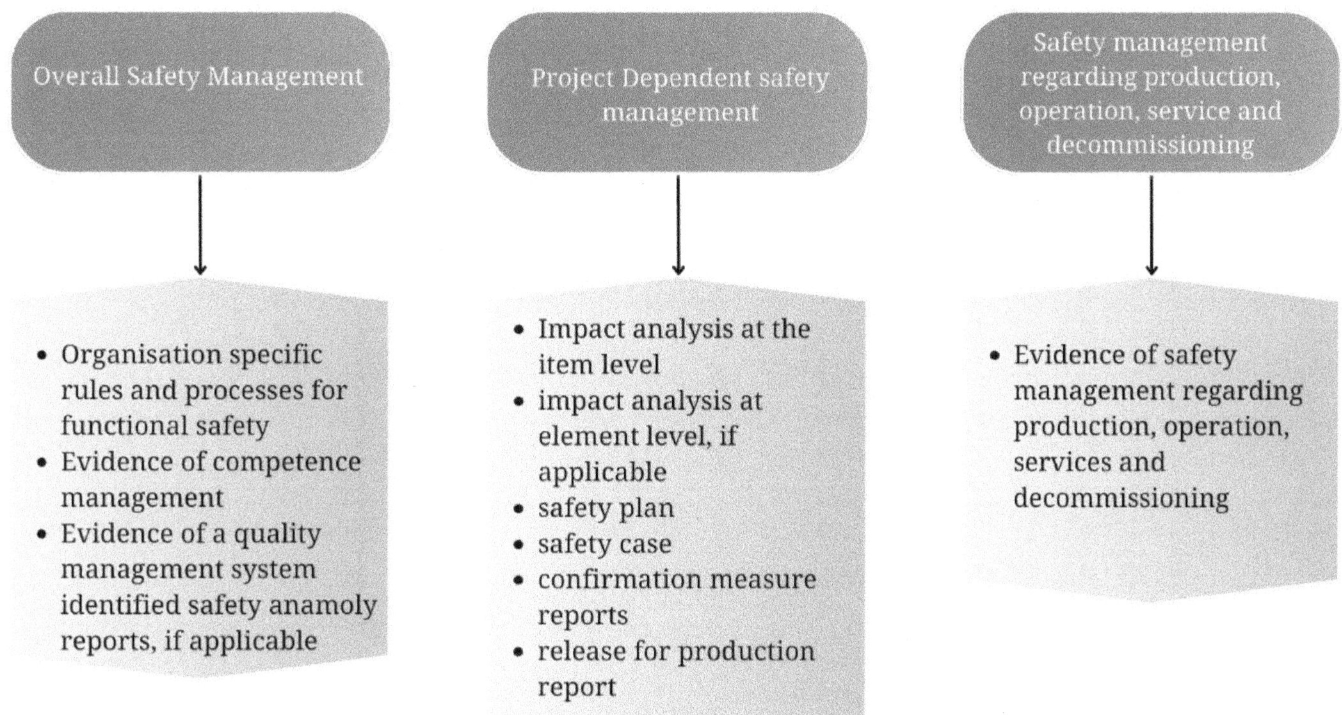

Two major work products that are performed during the project development are

Safety Plan – Covers overall Safety management activities, impact analysis and planning activities of the project

Safety case – Covers release for production approval and the result of all confirmation measures.

Evidence of safety management regarding production, operation, service and decommissioning will contain requirements that are provided by the functional safety manager that have to be followed during the production and maintenance process. These requirements will be integrated and implemented by the production incharge of the project. In addition, this document will contain the name of the responsible persons for production and maintenance.

Planning and Coordination of Safety activities - Safety Plan

Why is the Safety Plan Critical in Development of ISO 26262?

A product development team may opt for various different approaches to the System Development Life cycle. One of such proven approaches is Plan-Do-Check-Act, a general practice followed during project planning, especially in compliance verification scenarios.

PLAN– Development Interface Agreement, Safety Plans

DO– Concept documents, analysis documents, software codes

CHECK– any form of validation like MISRA C and other audits,

ACT– on preventive actions based on the derivations from Check stage.

With the advent of ISO 26262, the so-called security planning dimension has become an essential part of the project management plan (PLAN-Do-Check-Act). ISO 26262 requires organizations interested in applying security in the development of automotive software to adopt a strong safety culture.

The safety plan is prepared by the safety manager and shall be approved by the project manager. Safety Manager is responsible for maintaining the safety plan.

Required Inputs

- Customer Specification

- Item definition (System related information)

- Development Interface agreement (All Safety Work product Responsibility agreement with customer and supplier)

- Hazard Analysis Risk Analysis (ASIL and Safety goal)

- L1 planning (Target timeline information)

- Project team information (from project manager)

- ISO26262 V cycle and Internal V cycle

What should the Safety development Plan contain?

The Safety Development Plan would be the document that provides the complete project information apart from detailed technical requirements. The content of the Safety shall include

Safety Activities Timeline – This will be shared with safety engineers, for them to follow it during the delivery of safety workproducts of upcoming project phases. Each activity in the safety plan to have

Objective

- Dependencies on other confirmations if any

- Required resources for carrying out the activity

- Schedule with start and end date

- Prerequisites for each Activity

- Identification of the respective work product

Project Team Information – This shall include complete project team name and contact information including the Safety reviewer and assessor. Their roles and responsibilities in the project shall be clearly defined.

Escalation matrix – As we saw earlier, there will be an escalation matrix incase of any conflict of ideas faced in the duration of the project. This matrix shall be defined.

Input documents – The document received from customer and the project manager shall be recorded along with the storage location of the documents.

Project description – The Item description of the project to be included for a general understanding of the project

Customer Requirements – The safety requirements provided by customer or the requirements from HARA that are to be implemented in this particular project has to be documented along with their respective ASIL, Safe State and FTTI data.

Tools and Software used – The tools and resources required for the project shall be defined and their qualification process shall also be defined in the safety plan.

Tailoring of Safety Activities – List of tasks required as part of Safety Life Cycle according to ISO 26262 process. The applicable Safety activities shall be tailored by the Safety manager, and rationale shall be given for the safety activities that are not applicable. The timeline and review process that has to be followed for each safety activity shall be defined.

Safety Methodology followed for the project – The flowchart/methodology for the Safety activities for the project has to be defined. For example, the technical safety concept shall be used to define hardware and software requirements. It can also affect hardware-software integration documentation. The flow of safety requirements shall be defined from high-level to low-unit-level.

Traceability of requirements – The requirements shall be traceable by a tool like DOORS or Reqtify. The flowchart for Safety requirements defined by Safety methodology shall be confirmed with 100% coverage of high-level safety requirements.

Qualification of Hardware – The qualification process of Hardware shall be defined. Hardware components will be qualified based on ISO26262 – Part 8: Supporting Processes, Clause 13. All hardware electronic components must be ISO/TS16949 and AECQ100, AECQ101, AECQ200 compliant.

Qualification of Software – The software component qualification and any Component-Off-The-Shelf software components qualification process shall be defined. All components shall be qualified to the highest ASIL in the project.

Configuration Management – The configuration management shall assure: unique identification of product configuration (documentation and BOM); unique identification of each element of the configuration. The storage and maintenance of the project documentation shall be clearly defined.

Meetings with customer and suppliers – The frequency and duration of internal meetings, meetings with customers, meeting with suppliers shall be defined. The minutes of meetings for all the meetings shall be documented and maintained.

Issue tracking list – The issues identified during the progression of the project, right from the Item definition phase, have to be listed and the status has to be tracked. The location of this list has to be added in the Safety plan for record.

Review and assessment information – The reviewer, the independence of the reviewer, the assessor and their independence has to be clearly stated and should be in accordance with the highest ASIL rating in the project. The review and confirmation process for each safety activity to be clearly defined. For example, FMEDA has to undergo formal review and assessment. Whereas, for a development interface agreement document, a cross-read review is sufficient.

Document approval process – Each safety document has to undergo an approval process from the project team. The applicability for cross-read review with the project team, their responsibilities has to be defined in Safety Plan.

Change management process – There will be cases where the customer changes their requirements during the project. This will be considered as a change request and has to be handled with a defined process. The timeline in which the change can be integrated and delivered will be determined once a change is received. The responsible who will be involved in the change management and the process has to be defined in the Safety Plan.

Documenting the Evidence of Safety

Safety Case [2][3]

A safety case shall be developed, in accordance with the safety plan, in order to contribute an argument for the achievement of functional safety. The safety case should progressively compile the work products that are generated during the safety lifecycle to support the safety argument. In the case of a distributed development, the safety case of the item can be a combination of the safety cases of the customer and of the suppliers, which references evidence from the work products generated by the respective parties. Then the overall argument of the item is supported by arguments from all parties

To support progressive functional safety assessments the safety case can be released progressively as work products are generated to provide evidence for the safety arguments. Therefore, Safety case is not an activity to be performed towards the end of safety life cycle. The purpose of a safety case is to provide a clear, comprehensive and defensible argument, supported by evidence, that an item is free from unreasonable risk when operated in an intended context.

There are three principal elements of a safety case, namely:

- the safety goals and safety requirements

- the safety argument; and

- the ISO 26262 series of standards work products (i.e. the evidence)

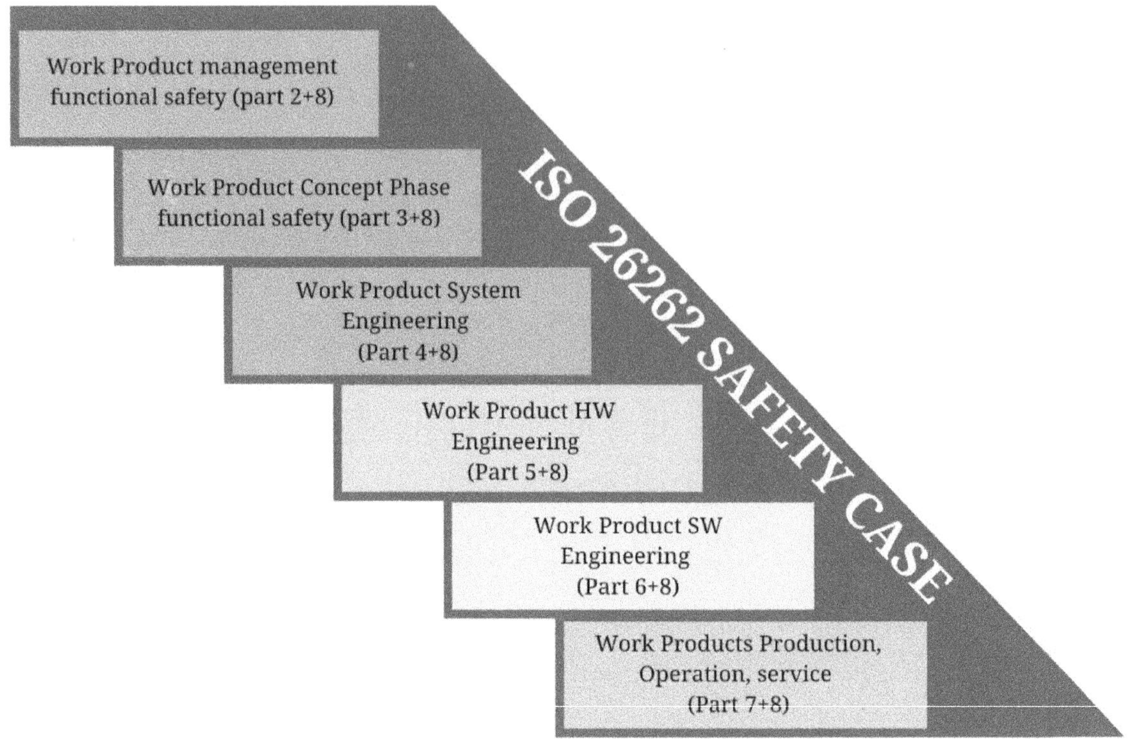

The main difficulty lies in the complexity of the safety case. As can be seen, many different (plans, specifications, reports, etc.) and various study materials are required. Required work items also arise at various stages of the project. At least two companies (customers and suppliers) are usually involved in the creation of data. A key element of a good safety case is project management. This management is presented in a compact but clear plan (safety plan, test plan, etc.). This compact traceability allows you to monitor projects at any time.

Another important issue is traceability at system/software/hardware level and environment. The main exhibits are requirements, architecture, source code, hardware schematics and tests. It is very important to have a strategy for device interfaces here. Traceability of two structures managed by different tools (like schema and requirements) requires some knowledge. If you then manage to find an approach on how to continuously add artifacts to the existing safety case, then you are on a very good way to develop a comprehensible, good and understandable safety case.

Safety Activity Management of a project as per ISO26262

The purpose of Part 2 of ISO26262[1] is to instruct how a project should be managed to make it compliant with ISO26262 standard. The Safety lifecycle provides us an overview on the phases a project will go through and the respective safety measures that will be taken for each phase.

The Safety Lifecycle

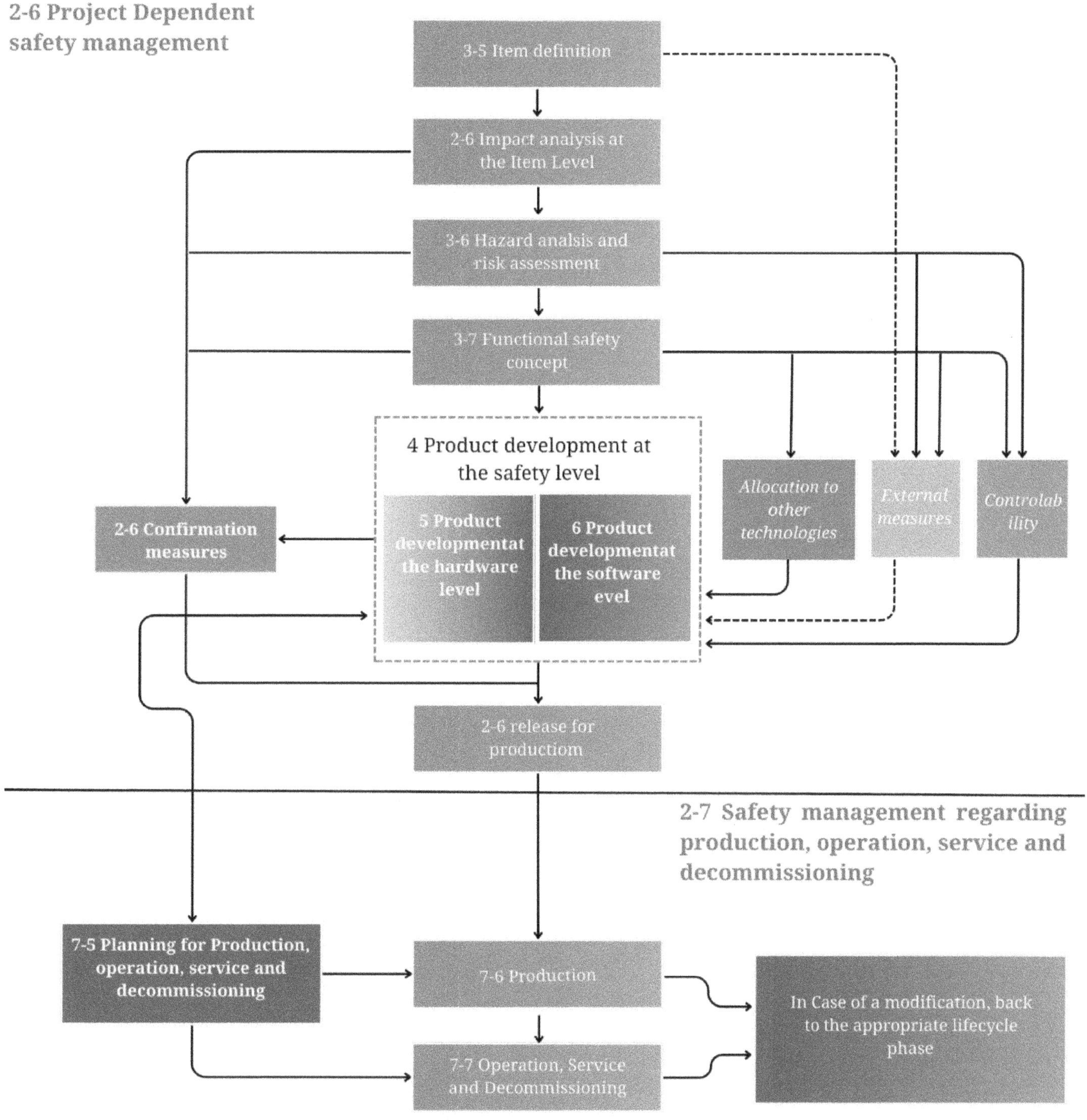

The key safety management tasks are to plan, coordinate and monitor functional safety-related activities. These management tasks apply to all phases of the safety lifecycle. As we saw earlier, the requirements for the management of functional safety given in this Part2 of ISO26262 are:

- overall safety management

- project dependent safety management, regarding the concept phase and the product development phases at the system, hardware and software level and

- safety management regarding production, operation, service and decommissioning

The planning of the safety activities regarding development is initiated at the concept phase and is refined as necessary through the product development phases (system, hardware and software) until the decision to release the item, or element, for production. Planning of activities related to production, operation, service and dismantling is initiated during the product development at the system level.

We will dive into detail for each phase and sub-phases as below

3-5 Item definition (a sub-phase of the concept phase)

The initiating task of the safety lifecycle is to develop a description of the item with regard to its functionality, interfaces, environmental conditions, legal requirements, known hazards, etc. The boundary of the item and its interfaces, as well as assumptions concerning other items, elements, or external measures are determined.

2-6 Impact analysis at the item level

An impact analysis is performed at the item level to determine whether the item is a new development, a modification of an existing item, or an existing item with a modified environment. If there are one or more modifications, the implications of the modifications on functional safety are analyzed.

2-6 Impact analysis at the element level

An impact analysis is performed at the element level when an existing element is reused, so as to evaluate whether the reused element is able to comply with the safety requirements allocated to that element, considering the operational context in which the element is reused.

3-6 Hazard Analysis and Risk Assessment (a sub-phase of the concept phase)

First, the hazard analysis and risk assessment estimates the probability of exposure, the controllability and the severity of the hazardous events with regard to the item. Together, these parameters determine the ASILs of the hazardous events and the safety goals for the item. The safety goals being the top level safety requirements for the item assigned with the ASILs determined for the corresponding hazardous events. The Human behavior considerations, including controllability and human response, in the hazard analysis and risk assessment, the functional safety concept and the technical safety concept, as well as the technical assumptions relevant for the ASIL classification are validated. In the next phase sequence, and subphases, detailed safety requirements are derived from the safety goals. A safety requirement inherits the ASIL of the corresponding safety goal, or receives the ASIL after decomposition in the case requirements decomposition with respect to ASIL tailoring has been applied.

3-7 Functional safety concept (a sub-phase of the concept phase):

Based on the safety goals, a functional safety concept is developed considering the preliminary architectural assumptions. It is developed by deriving functional safety requirements and by allocating them to the elements of the item. The functional safety requirements inherit the ASIL of the top level Safety goals.

The functional safety concept may also incorporate other technologies or rely on external measures. In those cases, the corresponding assumptions or expected behaviors are validated. The implementation of other technologies is outside the scope of the ISO 26262 series of standards and the implementation of the external measures is outside the scope of the item development in context of ISO 26262.

4 Product development at the system level

After clarifying the functional safety concept, the item is improved at the system level. The development process is based on the concept of the V model, the left side is the introduction of security regulations, design, design and implementation, and the right side is integration, identification and safety. The hardware-software interface is specified in step 4. The interface between hardware and software is changed during hardware and software design.

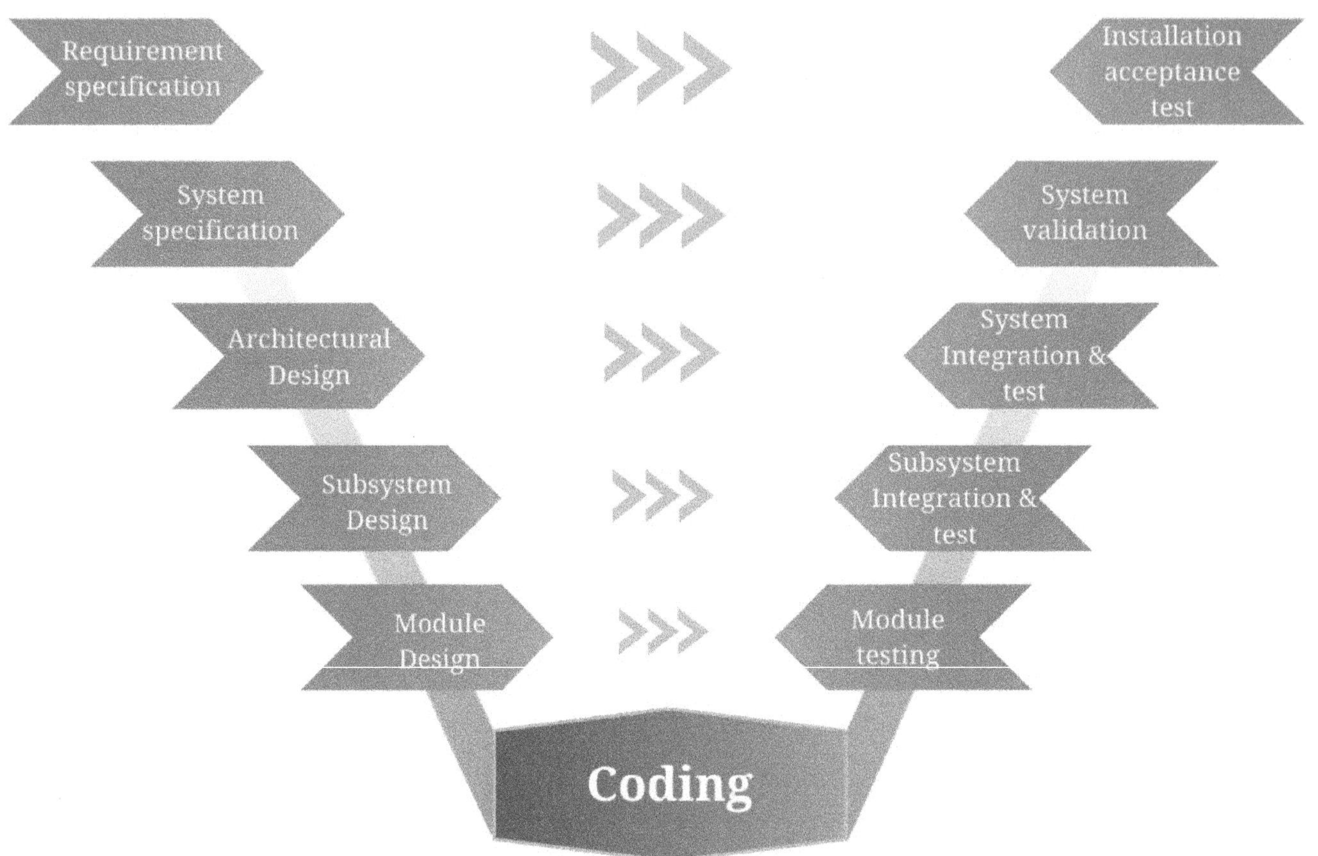

The system development incorporates safety validation tasks for activities occurring within other safety lifecycle phases, including:

- the technical assumptions relevant for the ASIL classification;

- the validation of the assumptions concerning human behaviour, including controllability and human response;

- the validation of the aspects of the functional safety concept that are implemented by other technologies; and

- the validation of the assumptions concerning the effectiveness and the performance of external measures.

5 Product development at the hardware level

Based on the system design specification, the hardware is developed. The hardware development process is based on the concept of a V-model with the specification of the hardware requirements and the hardware design and implementation on the left side and the hardware integration and verification on the right side.

6 Product development at the software level

Based on the system design specification, the software is developed. The software development process is based on the concept of a V-model with the specification of the software requirements and the software architectural design and implementation on the left side, and the software integration and the verification on the right side.

2-6 Confirmation measures

The confirmation measures are performed to judge the functional safety achieved by the item, or the contribution to the achievement of functional safety e.g. concerning the development of elements.

Controllability

In the hazard analysis and risk assessment, credit can be taken for the ability of the driver to avoid the specified harm, possibly supported by external measures. The assumptions regarding the controllability in the hazard analysis and risk assessment and the functional and technical safety concept are validated by specific confirmation measures.

External measures

The external measures refer to the measures outside the boundary of the item that reduce or mitigate the potential hazards resulting from malfunctioning behaviour of the item. External measures can include additional in-vehicle devices such as dynamic stability controllers or run-flat tyres, but also devices external to the vehicle, such as crash barriers or tunnel fire-fighting systems.

The assumptions regarding the external measures in the item definition, the hazard analysis and risk assessment and the functional and technical safety concept are validated by the confirmation measures.

Allocation to Other technologies

Other technologies (e.g. mechanical and hydraulic technologies) are those different from electrical and electronic technologies. These can be considered in the specification and allocation of safety requirements, or can be used as an external measure.

2-6 Release for production

The release for production formalizes the decision to release the item, or element, for production, considering the results of the safety lifecycle, including the results of the applicable confirmation measures.

7-5 Production, operation, service and decommissioning

The planning of this phase, and the specification of the associated requirements, starts during the product development at the system level and takes place in parallel with the system, hardware and software development.

Example: Assembly instructions for a camera lens in a Driver Monitoring System.

In a Driver Monitoring System, let's assume the output of the system regarding the driver alertness is safety relevant. Then the image of the driver captured is safety-critical. The camera assembly and lens play a critical role in capturing a proper image. These are mechanical items, which cannot be detected by the electronics incase of a defect. This will be a Safety-related requirement identified at system phase and then communicated to the production line. This phase ensures that we follow all such special safety requirements.

Here, we have safety-related special characteristics or requirements that improve the ability to produce the proper product, received from all previous phases. This phase addresses the processes, means and instructions to ensure functional safety regarding production, operation, service and decommissioning of the item or element.

The safety-related special characteristics and the development and management of instructions for the production, operation, service (maintenance and repair) and decommissioning of the item or element are considered.

Interpretation of Tables and Requirement applicability for an ASIL

In ISO26262, tables are used to define the requirements and confirmation methods needed for a particular clause and its level of recommendation for each ASIL. The different methods listed in a table contribute to the level of confidence in achieving compliance with the corresponding requirement.

Each method in a table is either:

a. a consecutive entry (marked by a sequence number in the leftmost column, e.g. 1, 2, 3), or

b. an alternative entry (marked by a number followed by a letter in the leftmost column, e.g. 2a, 2b, 2c).

For consecutive entries, all listed highly recommended and recommended methods in accordance with the ASIL apply. It is allowed to substitute a highly recommended or recommended method by others not listed in the table, in this case, a rationale shall be given describing why these comply with the corresponding requirement. If a rationale can be given to comply with the corresponding requirement without choosing all entries, a further rationale for omitted methods is not necessary.

For alternative entries, an appropriate combination of methods shall be applied in accordance with the ASIL indicated, independent of whether they are listed in the table or not. If methods are listed with different degrees of recommendation for an ASIL, the methods with the higher recommendation should be preferred. A rationale shall be given that the selected combination of methods or even a selected single method complies with the corresponding requirement.

Requirement applicability for an ASIL:

For each method, the degree of recommendation to use the corresponding method depends on the ASIL and is categorized as follows:

— "++" indicates that the method is highly recommended for the identified ASIL;

— "+" indicates that the method is recommended for the identified ASIL; and

— "o" indicates that the method has no recommendation for or against its usage for the identified ASI

Example for a consecutive entry:

Table 2 — Hardware design safety analysis

	Methods	ASIL			
		A	B	C	D
1	Deductive analysis	o	+	++	++
2	Inductive analysis	++	++	++	++

From the above example, both deductive and inductive analysis methods should be used for ASIL B, C, D. Rationale/justification shall be given if deductive analysis is not done for ASIL B as it is recommended(+). Highly recommended(++) methods shall be done mandatorily.

Example for alternative entry:

Table 11 — Methods for deriving test cases for software integration testing

	Methods	ASIL			
		A	B	C	D
1a	Analysis of requirements	++	++	++	++
1b	Generation and analysis of equivalence classes[a]	+	++	++	++
1c	Analysis of boundary values[b]	+	++	++	++
1d	Error guessing based on knowledge or experience[c]	+	+	+	+

In this table, one or more methods can be used to satisfy an ASIL. For example, while defining test cases for software integration testing for a ASIL B project, either one of the options from 1a, 1b, 1c, can be used as they are highly recommended(++) methods. A rationale/justification must be given if any highly recommended(++) is not used.

Chapter 4

ISO26262 Part 3 – Concept Phase

Topics to be covered

1. What is the Concept Phase?

2. Safety Work products involved in Concept Phase

3. Item Definition

4. Impact Analysis

5. Hazard Analysis and Risk Assessment (HARA)

6. Output of HARA - Safety Goal and ASIL

7. Functional Safety Concept

8. Case study

What is the Concept Phase?

ISO 26262 is an international standard for the functional safety of road vehicles. It provides guidelines and requirements for the development of safety-related systems in the automotive industry. Part 3 of ISO 26262 specifically focuses on the concept phase of the safety lifecycle.

The concept phase is the initial phase of the safety lifecycle and is crucial for the development of safe and reliable automotive systems. During this phase, the overall concept of the safety-related system is defined, including the system's architecture, functionality, and safety goals. The concept phase sets the foundation for subsequent phases, such as system design, implementation, verification, and production.

ISO 26262 Part 3 outlines the activities, techniques, and considerations that need to be addressed during the concept phase. It provides guidance on the following key aspects:

- Safety goals and requirements: Establishing the safety goals and requirements that the system must meet to ensure functional safety.

- Hazard analysis and risk assessment: Identifying potential hazards, assessing their risks, and determining the necessary safety measures to mitigate those risks.

- Functional safety concept: Developing a concept that ensures the functional safety of the system. This includes defining the system architecture, functional requirements, and necessary safety mechanisms.

- Safety requirements allocation: Allocating the safety requirements to the various system components and ensuring that the safety goals are met at each level of the system.

Item definition: Defining the system's boundaries and identifying the items that are safety-related.

Verification and validation planning: Planning for the verification and validation activities to ensure that the safety requirements are adequately verified and validated.

By following the guidelines provided in ISO 26262 Part 3, automotive manufacturers and suppliers can systematically address safety considerations during the concept phase of the development process. This helps in ensuring that safety is integrated into the design from the early stages and lays the groundwork for the subsequent phases of development.

Concept phase of the product is where we sketch a rough idea of the product, its functional requirements, operating environment, its interfaces with the vehicle etc.

In ISO26262, the concept phase is where the product development kick-off is done. In this phase, the product requirements - functional, operational, limits of interference, placement of the product in the vehicle - are defined. This phase will be the initial phase of the product development life-cycle.

Part 3 of ISO26262 deals with this kick-off or concept phase of the product development. It gives requirements that have to be satisfied during this phase in order to certify the product as ISO26262 compliant.

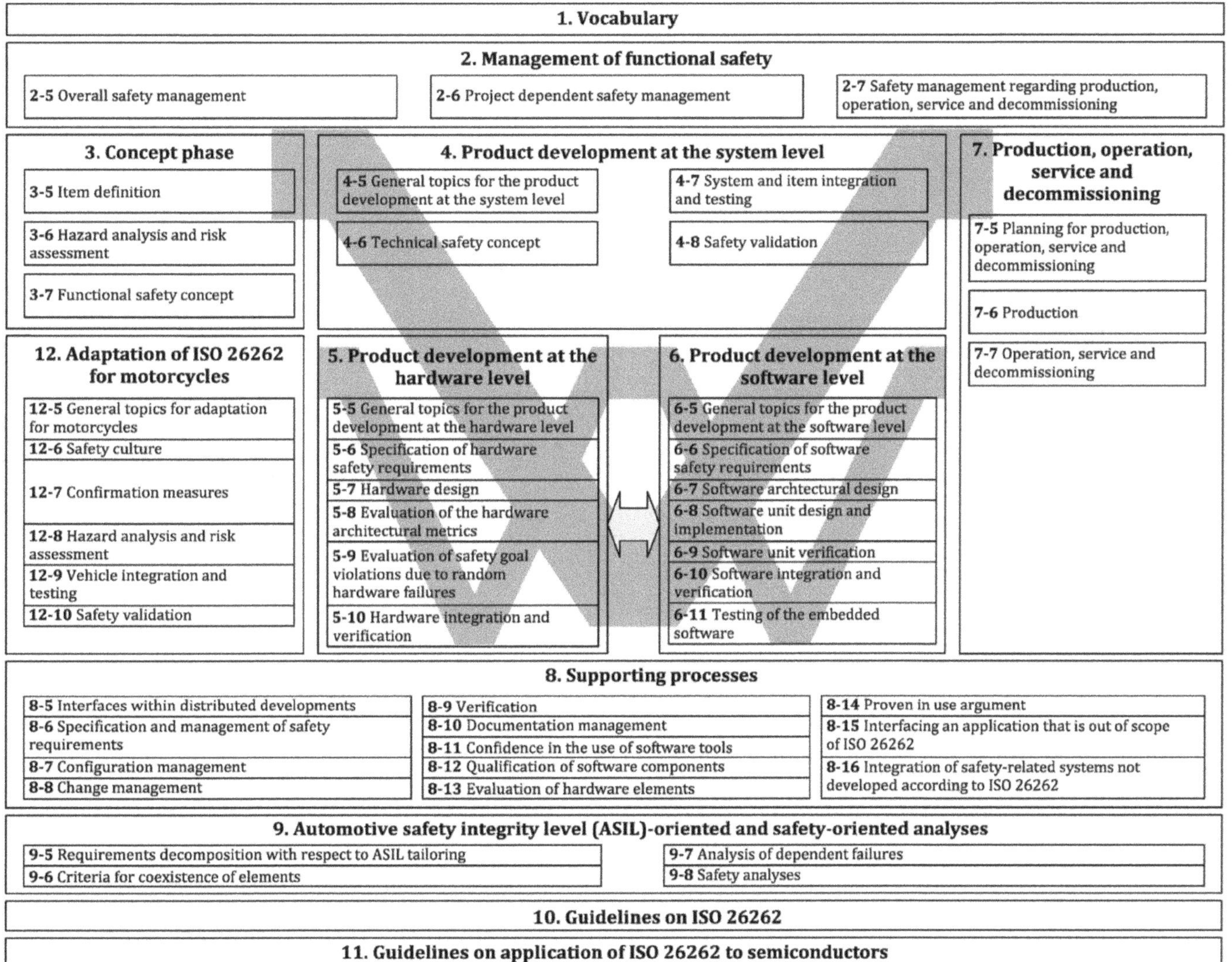

Safety Work products involved in Concept Phase

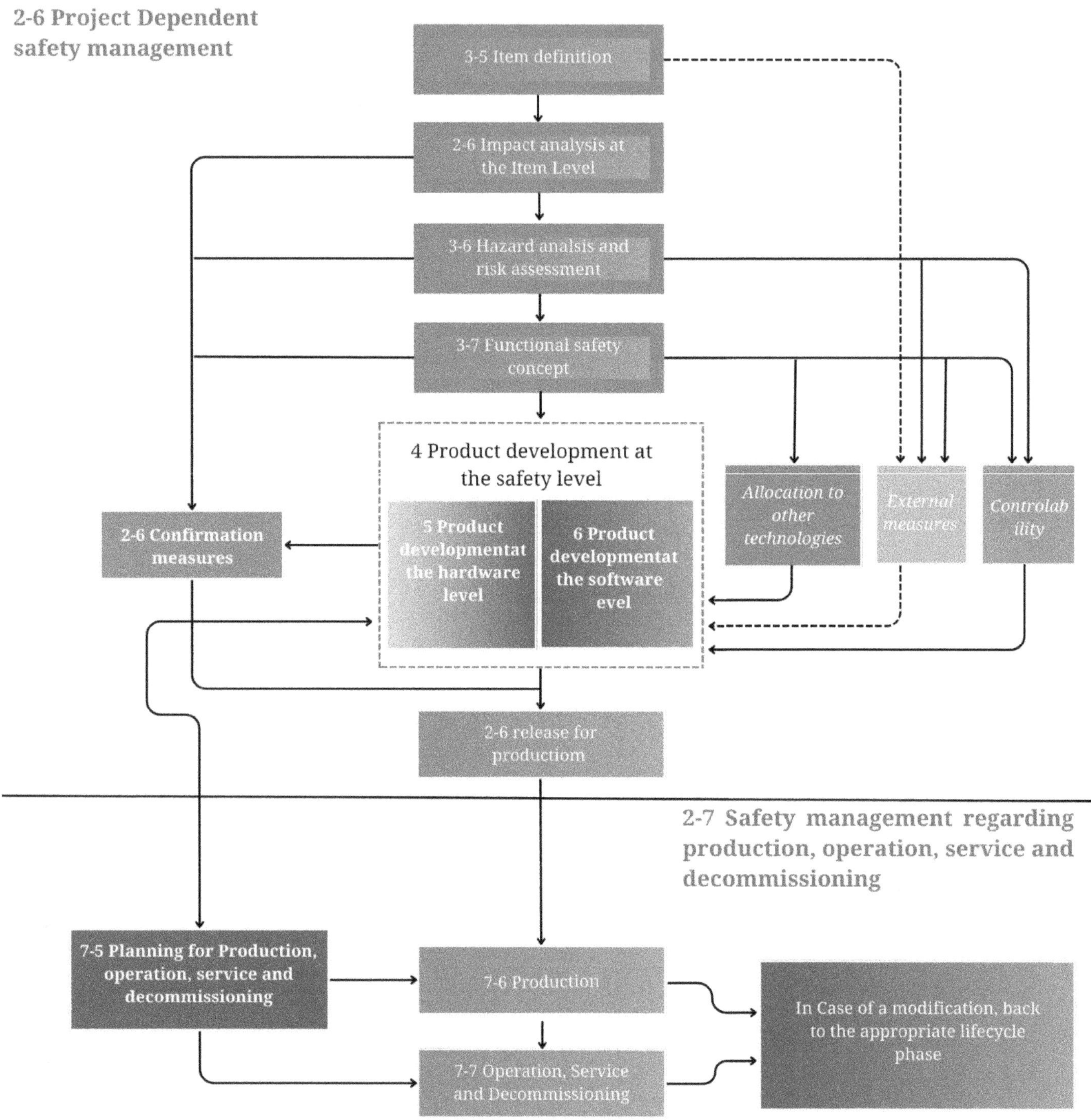

According to ISO26262 Part 3 [1] Annex A1, the safety work products that has to performed at concept phase, to comply the product with ISO26262, are as below:

- For Item definition - clause 5

- Item Definition - Functional Requirements and Specification of the product

- Impact analysis – In case of a previous reference project existing

- For Hazard analysis and risk assessment - clause 6

- Hazard analysis and risk assessment report

- Verification report of the hazard analysis and risk assessment

- For Functional Safety Concept - clause 7

- Functional Safety Concept

- Verification report of the functional Safety requirements

Let us look into each clause in detail in the upcoming sections

Item Definition

The term "item" denotes the subject of development, your product. These are one or more interacting electrical and/or electronic systems that implement the desired function.

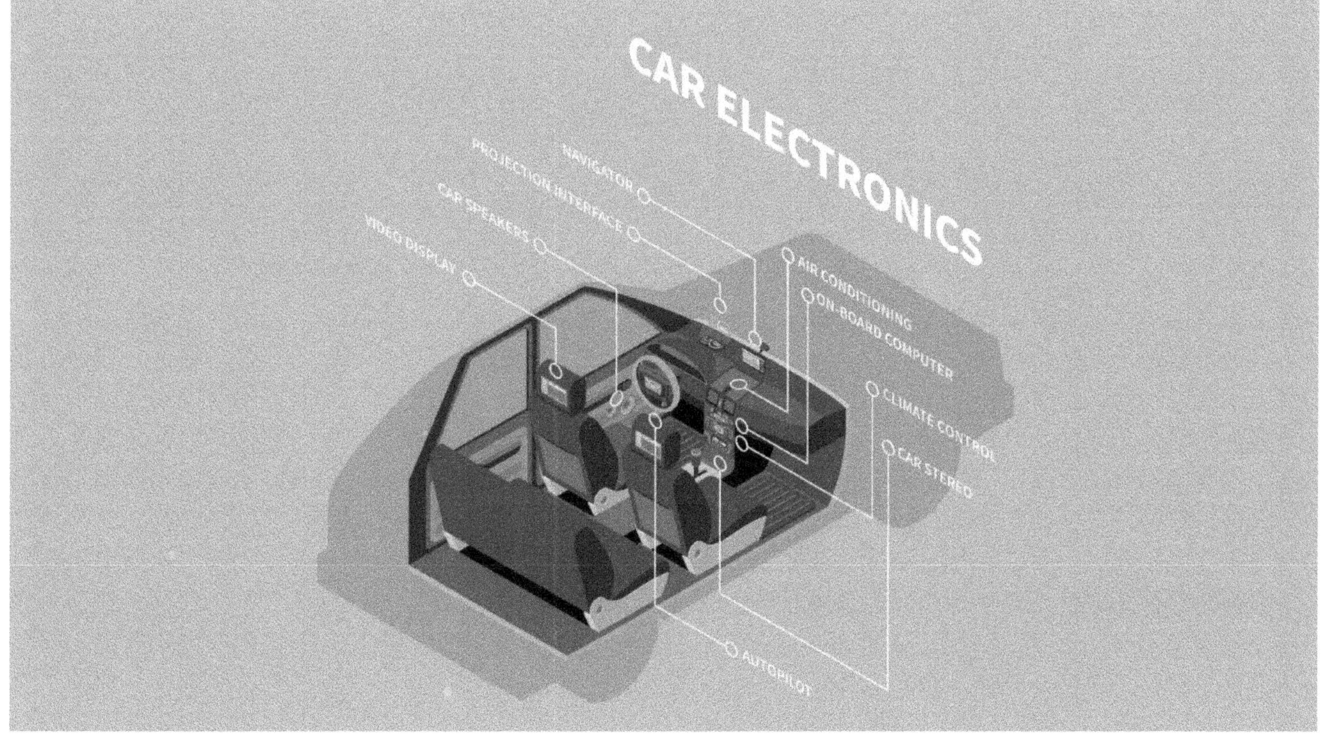

Examples of items are automatic cruise control systems, airbags or electrical components like car window mechanism. these simple electronic devices that can easily trap an arm or head. The development of the item means that different types of requirements and boundaries have to be brought together, such as operational requirements, documentation or the operation of the actuators included in the vehicle. It is also important that you agree on what lies outside the item, that is to know the boundary..

Impact Analysis

Impact Analysis shows how the lifecycle should be tuned, tailored and which safety activities are important. It is important to know whether the product was newly

designed, modified, or used in the past in a simply modified context, such as an airbag in a new type of car.

Now, in the safety lifecycle, this is more at the automaker and car level, but all suppliers need to have an impact analysis of their area of responsibility.

Example: An existing driver monitoring system was developed for a platform car and now the OEM has decided to make the same system for a truck. The changes in the process are as follows:

Design Component	Reuse/New	Comments
Item/System definition	Reuse	DMS function is the same
Hardware	New	Due to Truck design specifications - Height, camera angle, operating conditions
Software	Reuse	One image captured, same image processing algorithm software can be used

Hazard Analysis and Risk Assessment (HARA)

Risks to human life should be foreseeable in our activities. Note that this is usually a job done by the car company. As a risk assessment, more (sometimes less) needs to be done, both technically and organizationally.

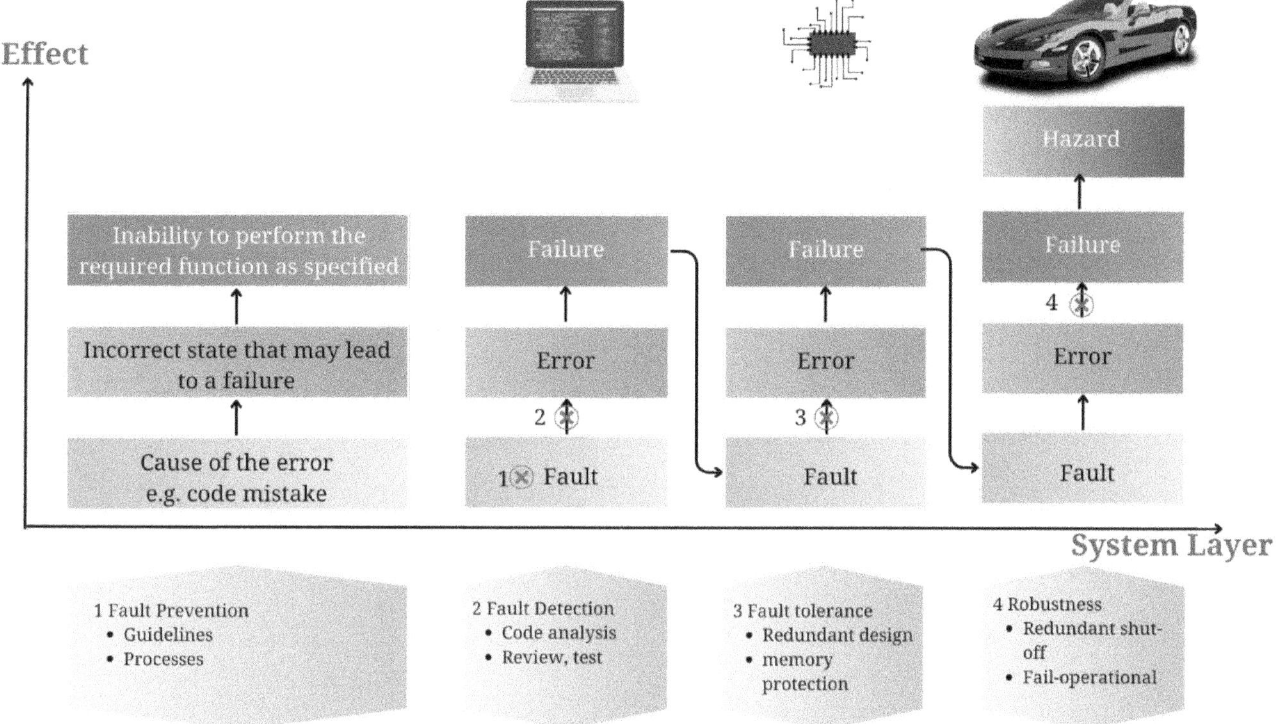

Let's consider an example of Hazard Analysis and Risk Assessment (HARA) in a vehicle:

Hazard: Unintended Acceleration

Hazard Identification: The hazard is identified as the vehicle experiencing unintended acceleration, where the vehicle starts accelerating without driver input.

Risk Assessment: The risk associated with unintended acceleration is assessed by considering factors such as severity, likelihood, and exposure. For example, unintended acceleration can lead to severe accidents, injuries, or even fatalities. The likelihood of occurrence may depend on various factors like vehicle design, electronic systems, and driver behavior. The exposure to this hazard can vary based on the driving conditions and the frequency of occurrence.

Risk Estimation: The level of risk is estimated by combining the severity, likelihood, and exposure factors. Based on the analysis, the risk level may be classified as low, medium, or high. For instance, if the severity is high, the likelihood is moderate, and the exposure is high due to a common driving scenario, the risk level may be determined as high.

Risk Mitigation: After identifying the risk, appropriate safety measures are determined to mitigate the risk of unintended acceleration. This may involve implementing various safety mechanisms such as redundant throttle control systems, brake override systems, sensor monitoring, and fail-safe mechanisms. Additionally, driver training and education regarding proper pedal usage and emergency procedures may also be considered as risk mitigation measures.

Verification and Validation: The effectiveness of the risk mitigation measures is verified and validated through various testing and analysis techniques. This ensures that the implemented safety mechanisms adequately address the identified hazard and reduce the associated risk to an acceptable level.

Hazard Analysis and Risk Assessment (HARA) is a continuous process throughout the development lifecycle of a vehicle, aiming to identify potential hazards, assess their risks, and implement appropriate safety measures to ensure functional safety.

The "automotive safety integrity level", or "ASIL", is then determined for the relevant hazardous events. This ASIL has a significant influence on development activities and the product.

Table 2 — Classes of probability of exposure regarding operational situations

	Class				
	E0	E1	E2	E3	E4
Description	Incredible	Very low probability	Low probability	Medium probability	High probability

Table 1 — Classes of severity

	Class			
	S0	S1	S2	S3
Description	No injuries	Light and moderate injuries	Severe and life-threatening injuries (survival probable)	Life-threatening injuries (survival uncertain), fatal injuries

Table 3 — Classes of controllability

	Class			
	C0	C1	C2	C3
Description	Controllable in general	Simply controllable	Normally controllable	Difficult to control or uncontrollable

To do this, you determine the "Severity" of harm, the probability of "Exposure" to the operational situation and the "Controllability", or ability to avoid harm. Once you have done that, you determine the ASIL.

$ASIL = f(S,C,E)$

ASIL is determined by the below table based on the Severity, Controllability and Exposure values determined by the scenario analysis.

Table 4 — ASIL determination

Severity class	Exposure class	Controllability class		
		C1	C2	C3
S1	E1	QM	QM	QM
	E2	QM	QM	QM
	E3	QM	QM	A
	E4	QM	A	B
S2	E1	QM	QM	QM
	E2	QM	QM	A
	E3	QM	A	B
	E4	A	B	C
S3	E1	QM	QM	A[a]
	E2	QM	A	B
	E3	A	B	C
	E4	B	C	D
a See 6.4.3.11.				

6.4.3.11 If several unlikely situations are combined that result in a lower probability of exposure than E1, QM may be argued for S3, C3 based on this combination.

For example, faulty steering by the lane-keeping assist system to the oncoming road could be classified as ASIL D, as serious injury could result. In contrast, it doesn't matter much if recognized traffic sign is displayed incorrectly because the traffic sign recognition system itself does not affect vehicle operation and the driver usually reacts appropriately. Once these tests are done, you can write your "Safety goal" for further development. Safety objectives are high-level safety requirements necessary to mitigate dangerous situations.

Output of HARA - Safety Goal and ASIL

A safety goal shall be determined for each hazardous event with an ASIL evaluated in the hazard analysis and risk assessment. If similar safety goals are determined, these may be combined together.

NOTE: Safety goals are high-level safety requirements for the item. They lead to the functional safety requirements needed to avoid an unreasonable risk for each hazardous event. Safety goals are expressed in terms of functional objectives. The determined ASIL for the hazardous event is assigned to the corresponding safety goal. If similar safety goals are combined together into a single one, the highest ASIL shall be assigned to the combined safety goal.

The safety goal can specify the fault tolerant time interval, or physical characteristics (e.g. a maximum level of unnecessary steering-wheel torque, maximum level of unnecessary acceleration) if they were relevant to the ASIL determination.

Assumptions used for, or resulting from the hazard analysis and risk assessment which are relevant for ASIL determination (if applicable, including hazardous events classified QM or with no ASIL assigned) shall be identified. These assumptions shall be validated in accordance with ISO 26262-4:2018,

Clause 8 for the integrated item.

NOTE Assumptions, if any, that are considered during the HARA include assumed actions of the driver or persons at risk and assumptions regarding external measures. Situation Analysis & Hazard Identification "Identify potential unintended behaviors of the item that could lead to a hazardous event."

- Vehicle Usage

- Environmental Conditions

- Foreseeable driver use and misuse

- Interaction between vehicle systems

Functional Safety Concept

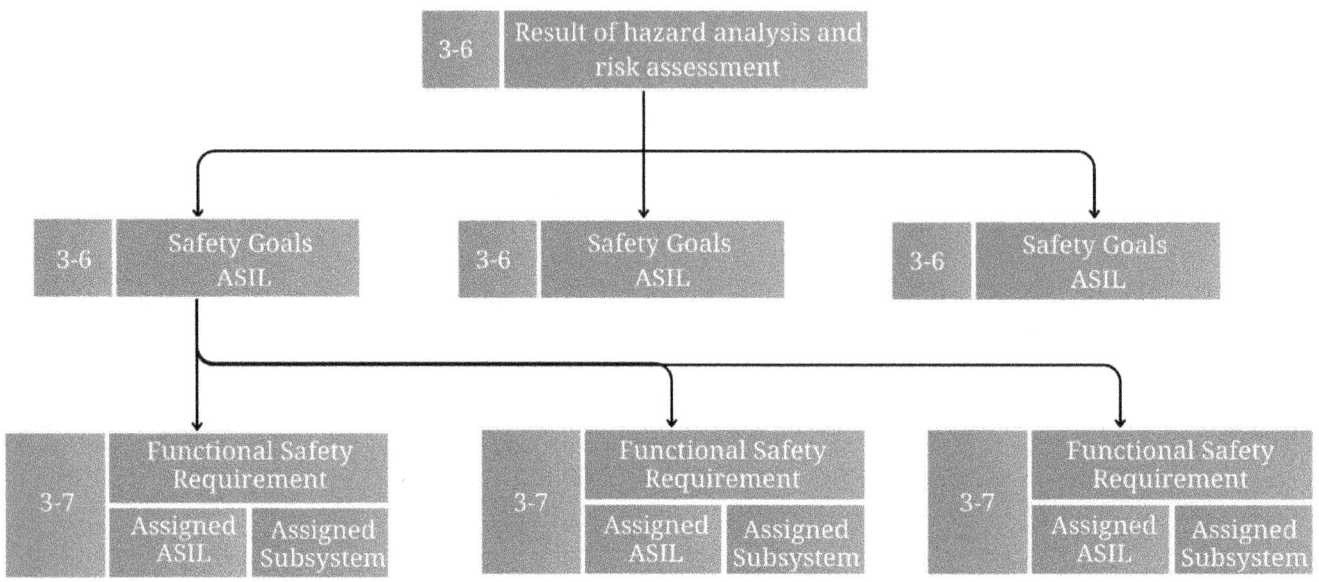

Functional Safety concept is composed of the Functional Safety Requirement

The fourth topic in the concept phase is the "functional safety concept". It is about deriving functional safety requirements ("FSRs") from the safety goals. At this point requirements for avoiding, detecting, and controlling faults are developed. A "safe state" is defined, into which the system changes in the event of an error, or which degraded state should be entered if the safe state cannot be reached immediately. Driver warnings are defined, to be displayed if an error.

Requirements must be assigned so that they get implemented in the system architecture or during external measures. The functional safety concept must be verified to determine it's suitability to adequate mitigation of the hazards.

Case study

Driver Monitoring System

Problem Statement: ABC transportation company was facing issues with driver safety. The company was facing an increasing number of incidents of drivers falling asleep or being distracted while driving, which was not only causing a risk to the drivers themselves but also to other vehicles on the road.

Detailed Analysis: To address this problem, the company decided to implement a driver monitoring system. They partnered with XYZ company, a provider of driver monitoring systems, and installed the technology in their entire fleet of commercial vehicles. The system was able to detect signs of driver fatigue or distraction, such as eye movements and steering patterns. It would then alert the driver through an in-vehicle alarm system, and also notify the company's central monitoring team.

The implementation process was relatively smooth. The entire system was installed within a month across the fleet. However, the teams faced challenge in effectively analyzing the data coming from the system, to identify patterns and proactively take measures based on the analysis.

Quantitative and Qualitative Data

After the implementation of the driver monitoring system, the company saw a significant improvement in driver safety. Specifically, the number of incidents where a driver fell asleep or was distracted while driving reduced by 55%. The company also reported a 15% decrease in overall vehicle collisions as a result of the early warning alerts system.

ABC transportation also found that the driver monitoring system provided an additional layer of peace of mind for their clients. The transportation company was able to communicate its focus on safety and provide assurance to clients that they take driver safety seriously.

Clear Results:The driver monitoring system has demonstrated tangible benefits, not only in improving driver safety and reducing the number of accidents but also in improving customer confidence and creating a competitive edge.

Compelling Narrative: The case study highlights that companies operating vehicles have a responsibility to ensure their drivers and other road users are safe. Driver monitoring systems indicate a firm's commitment to safety and can help enhance their reputation.

Through this partnership with XYZ company, ABC transportation was able to reduce the number of accidents caused by driver fatigue and distraction, thereby making the roads safer. The stories of some drivers who recorded unprecedented improvements in their driving behavior after the system installation was indeed fascinating.

R
Driver ID: James
Eye openness: 85%
Blinks per minute: 16
Gaze vector: 10°, -7°
Pupil dilation: 35%
Attentiveness: YES
Seatbelt: YES
Drowsiness: NO
Cellphone: NO
Smoking: NO
Pitch:
Roll: 0
Yaw: 15

HARA

Potential Hazard	Projected Effects	Associated Risks	Possible Controls
Loss of Power	Severe Collision	Critical	Dual Power System (including battery, wiring, and connectors)
Loss of Communication	Severe Collision	Critical	Dual Communication System (topology and tapping need to be determined)
Loss of Steering	Severe Collision	Critical	• Backup System • Reduced Functionality Redundant System • Steer by Braking Active Safety System
Loss of Braking	Severe Collision	Critical	• Backup System • Reduced Functionality Redundant System • Brake by Steering Active Safety System
Loss of Electronic Throttle	Severe Collision	Critical	• Backup System • Reduced Functionality Redundant System
Loss of Actuator(s)	Severe Collision	Critical	• Backup Actuator(s) • Degraded Performance Actuator(s)
Loss of Sensor(s) (recording driver commands)	Severe Collision	Critical	• Backup Sensor(s) • Degraded Performance Sensor(s)
Loss of Sensors (in feedback loops)	Driver Interaction, Driver Discomfort	Low to Moderate	Open Loop Control
Loss of Active Safety	Driver Discomfort	Low	None (driver required to drive more carefully)

FSC

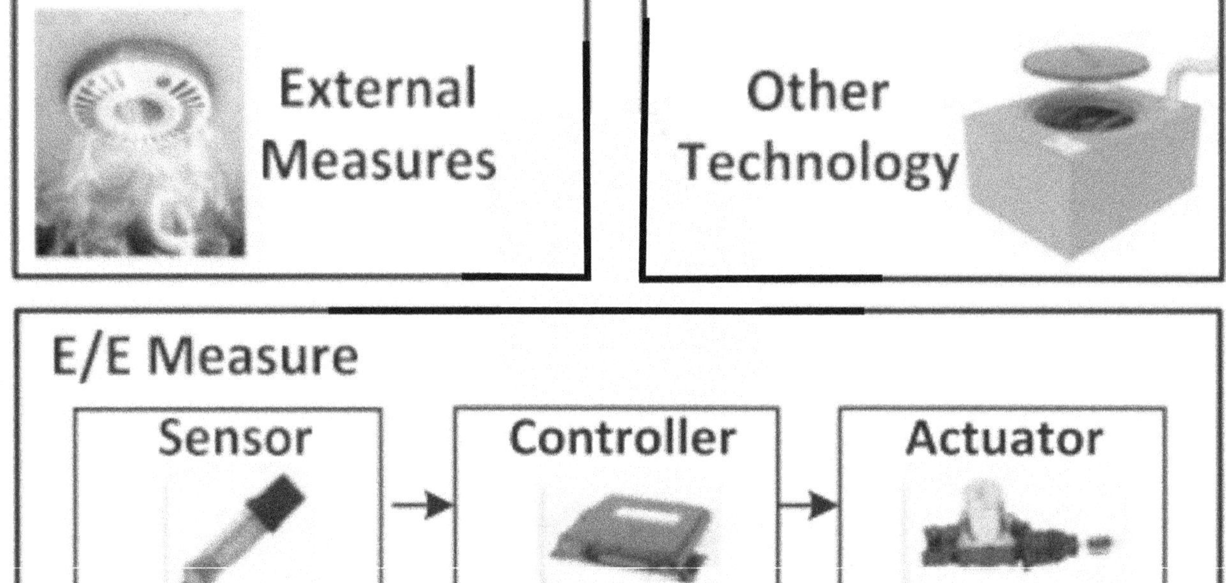
Functional Safety Concept for Safety Goal X
External Measures
Other Technology
E/E Measure
Sensor
Controller
Actuator

Chapter 5

Part 4 – Technical Safety Concept

Topics to be covered

- What is the Technical Safety Concept?

- What is its role in ISO26262?

- Inputs required for Technical Safety Concept definition

- How are the requirements defined?

- How is it confirmed to meet the high-level requirements

- Requirements formulated by ISO26262 Part 5

- Case Study

Technical Safety Concept

Technical Safety Concept is a formal document that defines the Technical Safety Requirements that are mandatory to be verified and validated in the development process of the product to comply with ISO26262 certification to meet an ASIL qualification.

Technical Safety Requirements (hereafter mentioned as TSRs) are preventive measures which counters a hardware and software faults to prevent the system from becoming a potential hazard during system failure.

The Role of Technical Safety Concept in ISO26262

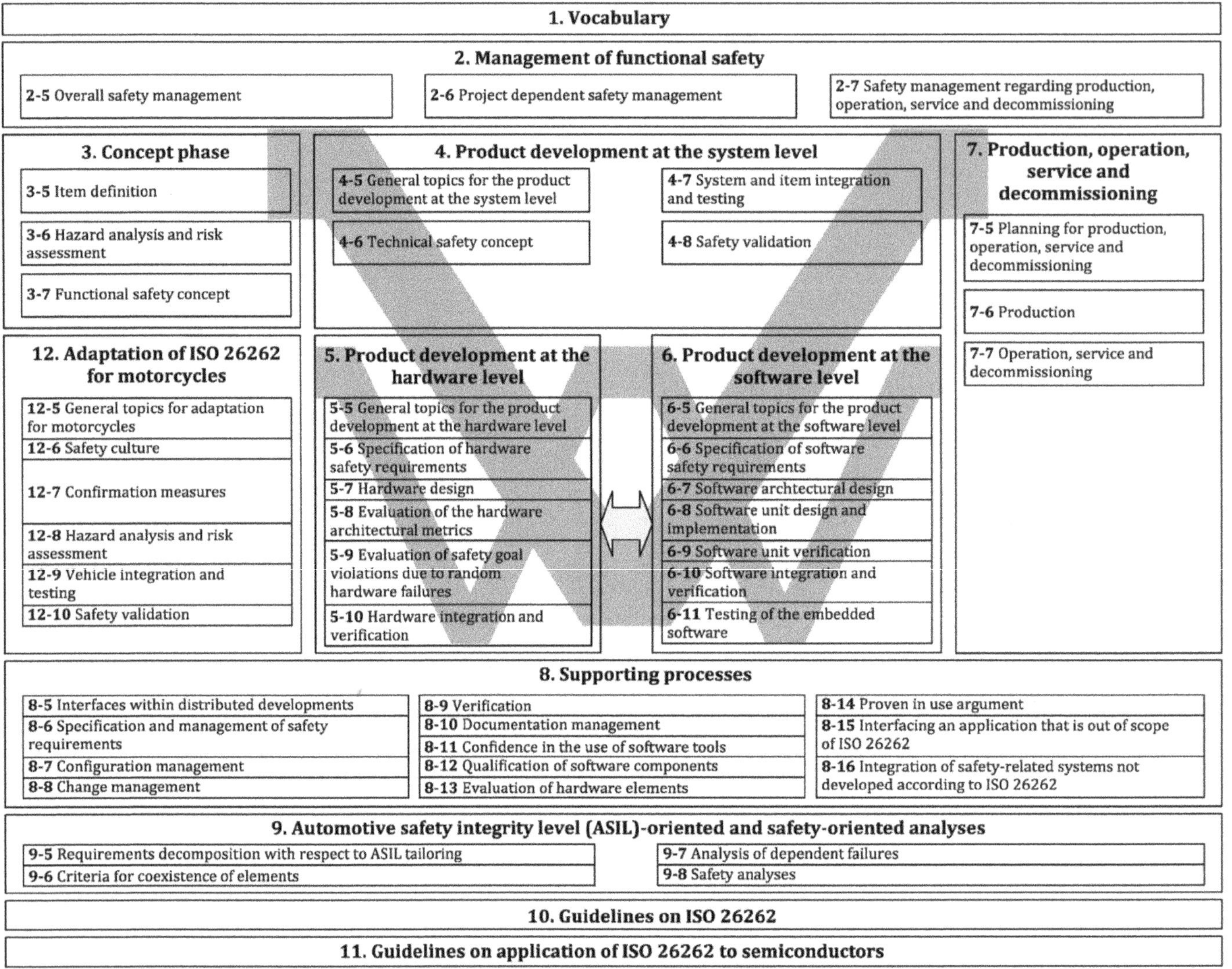

Technical Safety Concept is mandated by ISO26262 in Part 4 - the System level product development [1].

As Shown in the above flowchart, the 4-6 Technical Safety Concept would be the requirement document at System level that tracks the requirements defined at the Concept phase and refines the requirements so that each requirement can be assigned to Part5 - Hardware and Part-6 Software development phases of the project.

The requirements defined shall be atomic and assigned to each department, hence clearly stating the list of requirements that has to be followed at the next stages of the product development by each team.

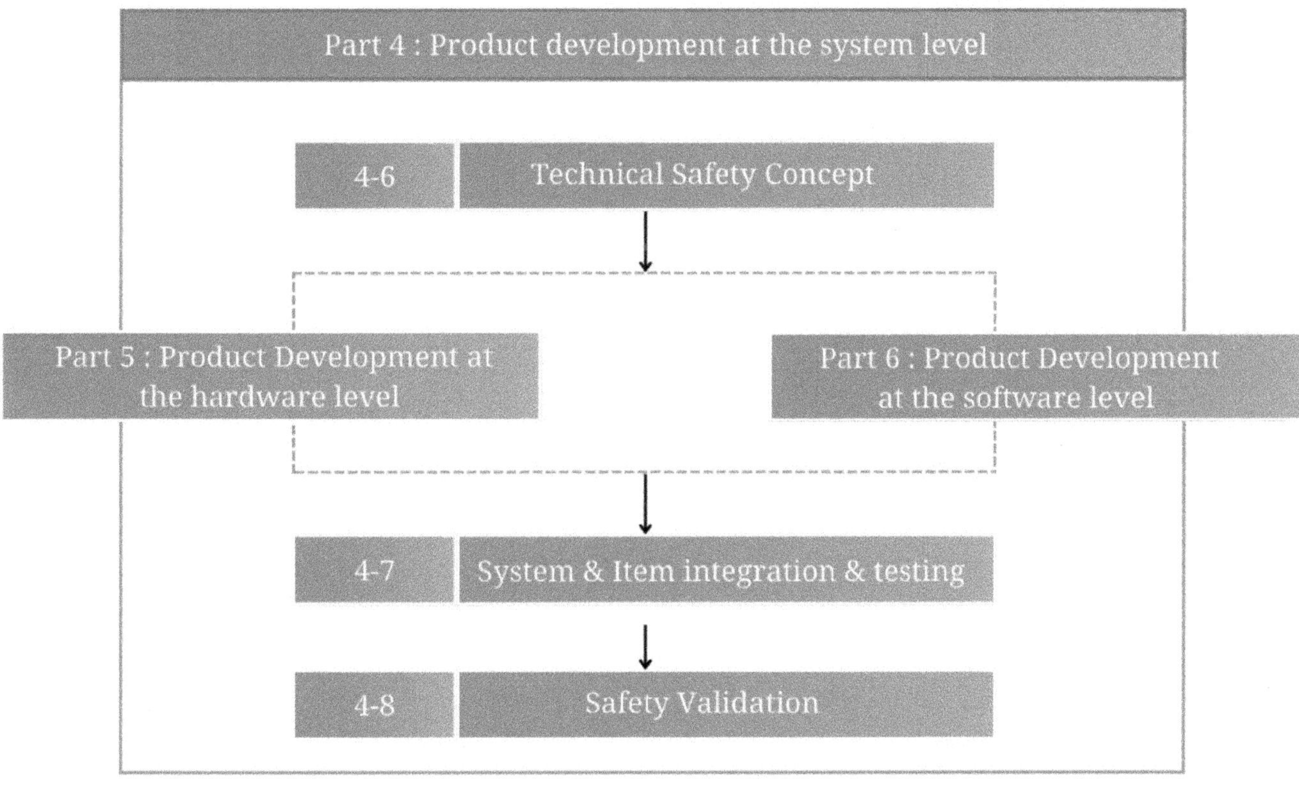

The TSRs defined in this document would also be the requirements against which the system would be tested in the System Integration and testing phase. So each TSR defined shall be testable and shall be validated for conformance of the product to follow ISO26262 standard development.

The Prerequisites/Inputs for Technical Safety Concept in ISO26262

Hazards identified at Concept level

The Hazards identified at the Hazard Analysis and Risk Assessment (HARA) performed at the Concept phase would be the main input base for the Technical Safety concept definition. The HARA should confirm the system's function has been evaluated for all possible environments and use cases of the product to define all the possible undesired events of all the functions defined for the product under analysis. The events shall be evaluated with maximum Severity, Exposure and Controllability to define respective ASIL level for each Safety goal and Undesired event.

Safety goals defined with ASIL and FTTI

The output of the HARA at concept phase gives the list of Safety Goals that have to be confirmed by the product. The Safety goals would be the requirement that prevent the system from leading to an undesired event. The Safety Goals shall be defined with Fault Tolerant Time Interval (FTTI) and ASIL level.

The Fault tolerant time interval is the interval in which the system will tolerate the hardware or software failure before violating the Safety goal.

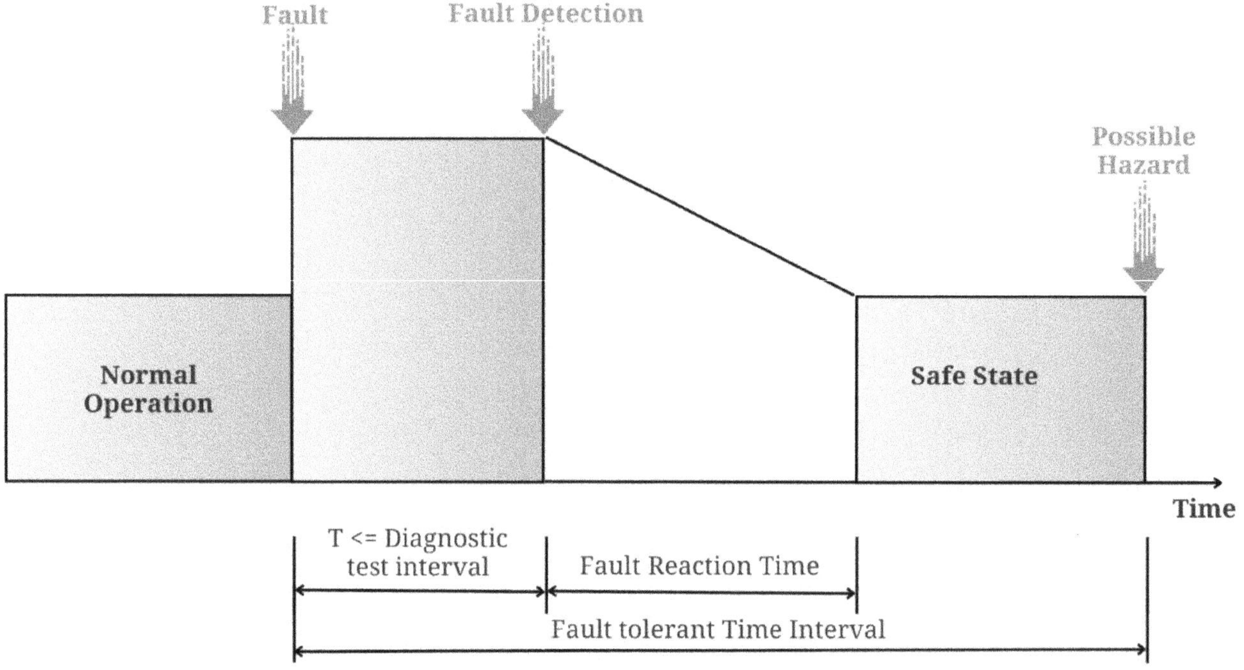

Each TSR defined shall take the upstream Safety goal's FTTI and ASIL level. If one TSR is applicable for two or more Safety goals, the minimum FTTI and maximum ASIL level shall be applied for the TSR.

During the definition of TSRs, the time taken to detect a fault and the time taken to reach the Safe state should be defined to be within the FTTI, so that we reach the Safe state in the event of failure before causing a hazard.

Safe State

The Safe state is one more output of HARA. This defines the state which the system can reach in the event of a hardware or software fault. The TSRs should be defined to detect the fault and also to define which safe state to be reached in event of which failure.

Hence the Safe State definition for each function would be an input for Technical Safety Concept definition.

Basic HW architecture

Functional Safety concept if available

How are the requirements defined?

As per ISO26262 Part 4 definition, the TSRs when defined in the technical safety concept shall consider the following topics:

1. The TSRs shall clearly define the configuration of the system i.e., the mode of the system shall be clearly defined for each TSR.

   Eg: In Normal Mode and Low power mode, the Microcontroller MC1 shall detect the loss of communication in the LIN interface (LIN1)

2. The TSRs shall clearly state the stimulus of the system when the failure that can lead to an undesired event is detected.

   Eg: When loss of communication on LIN1 is detected, Mircontroller MC1 shall command the RED led ON indicator with FTTI1.

3. TSR shall be defined to prevent all possible failure modes of intended functionality. These preventive measures are known as Safety mechanisms. All required safety mechanisms to prevent internal critical failures of each block inside the system/ component shall be defined as TSR.

   Eg: The LIN message "DoorStatus" shall be prevented with CRC mechanism to detect erroneous safety data in the message. If erroneous CRC is detected, the microcontroller shall discard the frame.

4. TSR shall define that the FMEDA shall be performed for metric conformance. It shall also clearly state the metrics that have to be met.

   Eg: The ASIL B function shall satisfy FMEDA analysis with below metrics:

   SPFM > 90% LPFM >6% PMHF < 100 FIT

   Note: The metrics levels are provided by ISO26262 and shall be followed unless defined otherwise by the customer.

5. TSR shall define Freedom from Interference requirements when multiple functions with different ASIL levels are achieved with the same component.

    Eg: The memory space of the microcontroller shall be separated for ASIL A and ASIL B variables. Any error in ASIL A variables shall not affect the ASIL B variables.

6. TSRs shall clearly state the Bottom-up requirements that are to be achieved by other components in the System/Vehicle to confirm safe operation of the component under analysis.

    Eg: The BCM module shall implement CRC and timeout safety mechanism for the LIN interface communication with the Doors Control Module. Here, the door control module would be our scope of analysis, but for safe operation of this module, we are giving a requirement that has to be implemented in an external module (BCM).

Conformance to meet the high level requirements

The Requirements defined will be considered to be enough for a particular ASIL level by performing FMEDA analysis

FMEDA analysis performs the below:

- Protects each local potential failure with Safety mechanism (Eg: Power supply supervisor for Undervoltage/Overvoltage failure)

- Calculate SPFM and LPFM using Diagnostic coverage from ISO26262 Part 5 Annex D

- The FMEDA analysis to define the SPFM, LPFM will be detailed in the Part 5 - Hardware Metric Analysis chapter of the course.

Requirements formulated by ISO26262 Part 5

The requirements shall be defined for each Hardware random failure. The ISO26262 Part 5 Annex D [2] defines what are the possible failures of a Hardware part and the Safety Mechanisms that can be defined to prevent the respective failures.

The higher the ASIL requirement, the safety mechanisms with higher diagnostic coverage shall be implemented, as we can note that higher the ASIL the higher the metric values to be met by the product safety design.

As defined by ISO26262 Part 5 [2], below are the metrics to be met by the design for respective ASIL levels.

Table 4 — Possible source for the derivation of the target "single-point fault metric" value

	ASIL B	ASIL C	ASIL D
Single-point fault metric	≥90 %	≥97 %	≥99 %

Table 5 — Possible source for the derivation of the target "latent-fault metric" value

	ASIL B	ASIL C	ASIL D
Latent-fault metric	≥60 %	≥80 %	≥90 %

Table 6 — Possible source for the derivation of the random hardware failure target values

ASIL	Random hardware failure target values
D	$<10^{-8}\ h^{-1}$
C	$<10^{-7}\ h^{-1}$
B	$<10^{-7}\ h^{-1}$
NOTE The quantitative target values described in this table can be tailored as specified in 4.2 to fit specific uses of the item (e.g. if the item is able to violate the safety goal for durations longer than the typical use of a passenger car).	

In the next few sections, we will see a few examples on how the guide to define Safety mechanisms are provided in the standard document ISO26262 Part 5 [2] Annex D.

TSRs for Microcontroller/Microprocessor

Table D.4 — Processing units

Safety mechanism/ measure	See overview of techniques	Typical diagnostic coverage considered achievable	Notes
Self-test by software: limited number of patterns (one channel)	D.2.3.1	Medium	Depends on the quality of the self-test
Self-test by software cross exchange between two independent units	D.2.3.3	Medium	Depends on the quality of the self-test
Self-test supported by hardware (one-channel)	D.2.3.2	Medium	Depends on the quality of the self-test
Software diversified redundancy (one hardware channel)	D.2.3.4	High	Depends on the quality of the diversification. Common mode failures can reduce diagnostic coverage
Reciprocal comparison by software	D.2.3.5	High	Depends on the quality of the comparison
HW redundancy (e.g. dual core lockstep, asymmetric redundancy, coded processing)	D.2.3.6	High	It depends on the quality of redundancy. Common mode failures can reduce diagnostic coverage
Configuration register test	D.2.3.7	High	Configuration registers only
Stack over/under flow Detection	D.2.3.8	Low	Stack boundary test only
Integrated hardware consistency monitoring	D.2.3.9	High	Coverage for illegal hardware exceptions only

NOTE   This table deals only with safety mechanisms dedicated to processing units. General techniques like one based on data comparison (see D.2.1.2) are also able to detect failures of electrical elements but are not integrated in this table (already included in Table D.2 — E/E Systems).

Some of the Safety mechanisms that can be defined for MCUs are:

ECC/CRC for RAM, ROM - prevents failure by memory corruption

Self test of Core - prevents failure by Core stuck or corruption

Watchdog - Internal/External - prevents failure by wrong program sequence, SW program stuck, etc.

Self test of Peripherals like ADC, CAN - prevents failure by corruption or loss of safety-relevant frames sent in the communication interface.

Self test of memories - prevents failure by Stack overflow, configuration register corruption.

Core lockstep - Higher ASILs

Redundant memory storage - Higher ASILs

TSRs for Communication

Table D.6 — Communication bus (serial, parallel)

Safety mechanism/ measure	See overview of techniques	Typical diagnostic coverage considered achievable	Notes
One-bit hardware redundancy	D.2.5.1	Low	—
Multi-bit hardware redundancy	D.2.5.2	Medium	—
Read back of sent message	D.2.5.9	Medium	—
Complete hardware redundancy	D.2.5.3	High	Common mode failures can reduce diagnostic coverage
Inspection using test patterns	D.2.5.4	High	—
Transmission redun-dancy	D.2.5.5	Medium	Depends on type of redundancy. Effective only against transient faults
Information redun-dancy	D.2.5.6	Medium	Depends on type of redundancy
Frame counter	D.2.5.7	Medium	—
Timeout monitoring	D.2.5.8	Medium	—
Combination of information redundancy, frame counter and timeout monitoring	D.2.5.6, D.2.5.7 and D.2.5.8	High	For systems without hardware redundancy or test patterns, high coverage can be claimed for the combination of these safety mechanisms

The above table gives us all possible safety mechanisms that can be added for any communication interface. Some of the commonly used Safety mechanisms that can be defined for MCUs are:

Timeout for periodic communication - prevent loss of frames

CRC for each frame - for corruption of data

Alive counter - for preventing frozen frames

Redundant communication interface - used to meet higher ASILs for ASIL decomposition purposes.

TSRs for Sensors

Table D.9 — Sensors

Safety mechanism/ measure	See overview of techniques	Typical diagnostic coverage considered achievable	Notes
Failure detection by on-line monitoring	D.2.1.1	Low	Depends on diagnostic coverage of failure detection
Test pattern	D.2.4.1	High	—
Input comparison/ voting (1oo2, 2oo3 or better redundancy)	D.2.4.5	High	Only if dataflow changes within diagnostic test interval
Sensor valid range	D.2.8.1	Low	Detects shorts to ground or power and some open circuits
Sensor correlation	D.2.8.2	High	Detects in range failures
Sensor rationality check	D.2.8.3	Medium	—

Similarly, the above table is provided by the ISO standard for sensor units. Few examples of safety mechanisms used are:

Self-Test using predetermined value - eg: connecting passive capacitor for any one channel and using other channel for actual capacitive sensing

Sensor range check - Detects short circuit or open circuit of sensor line which is out of usual range of the sensed value

Multiple redundant sensor with independence - used for Higher ASILs for ASIL decomposition purposes.

TSRs for Power Supplies

Table D.7 — Power supply

Safety mechanism/ measure	See overview of techniques	Typical diagnostic coverage considered achievable	Notes
Voltage or current control (input)	D.2.6.1	Low	—
Voltage or current control (output)	D.2.6.2	High	—

The above table provides the safety mechanism for power supply. The power supply is prone to have undervoltage, over voltage and loss of power supply failures. Hence, monitoring the power supply by a smart device that can move the system to safe state in event of any of those failures would be the safety mechanism define. Few example of such monitoring mechanisms are:

- Undervoltage monitoring using power supervisor circuit

- Monitoring for loss of supply by using MCU that is powered by another power supply.

- Overvoltage monitoring using internal voltage monitors or by ADC in MCU. This will mostly be used for lower ASILs as it has dependency that the MCU will be powered by the same power supply.

- The power supply monitoring supervisory circuit will be implemented as independent circuit for higher ASILs to confirm freedom from interference of the failure and the preventive measure.

Case Study

We will see an example of the Rear Occupant Alert system announced by Hyundai [3].

According to the statistics, 38 children on average have died from being stuck in hot vehicles every year since 1991. As it is hard to understand if someone is still in the back seat wif it's a child sleeping in a big car seat under the blanket. U.S. child safety and accident prevention organization, 39 children die of heatstroke in a locked car each year. Hyundai Motor Group provides a safety feature that can help prevent such unfortunate occurrences: Rear Occupant Alert (ROA).

The system consists of the radar sensor module hidden in the ceiling communicating via Controller Area Network(CAN) to the Integrated Body-control Unit(IBU) that generally controls the whole system of a car.

This system employs, Ultrasonic sensor detects movement in the rear seats and informs to IBU. IBU commands the Horn sounds, lights flash and a Blue Link alert message is sent to the driver's smartphone if movement is detected

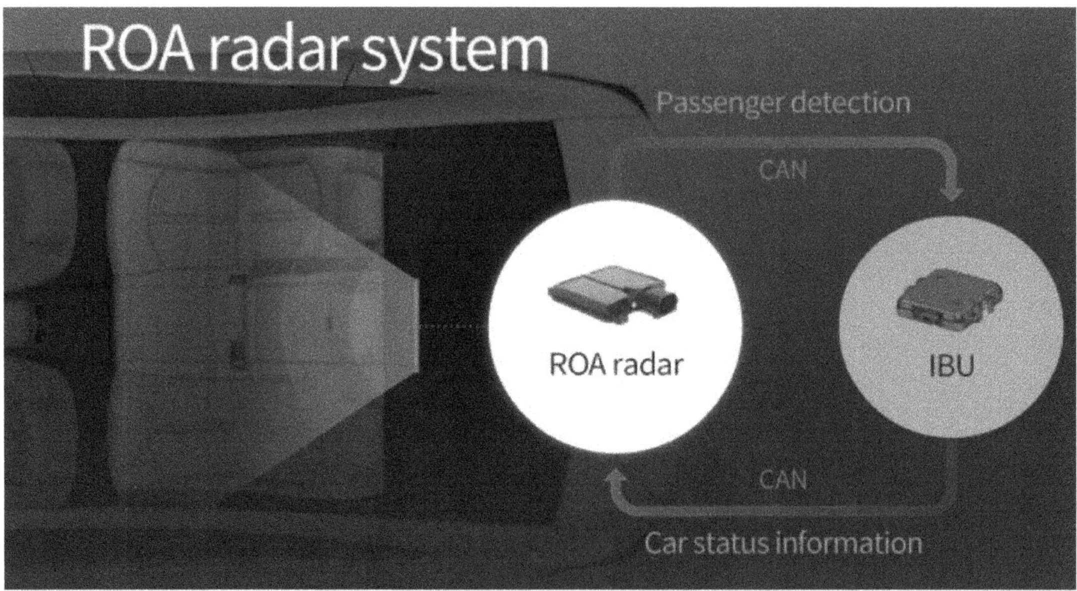

Let us dive deeper into the Ultrasonic sensor component. Based on the System level understanding, the component is expected to perform following functions:

- Sense the rear cabin for the car for possible live occupants.

- Communicate with the IBU regarding the rear-occupant status via CAN interface.

Based on these functional requirements, let us arrive with a suitable Hardware Block Diagram for the unit:

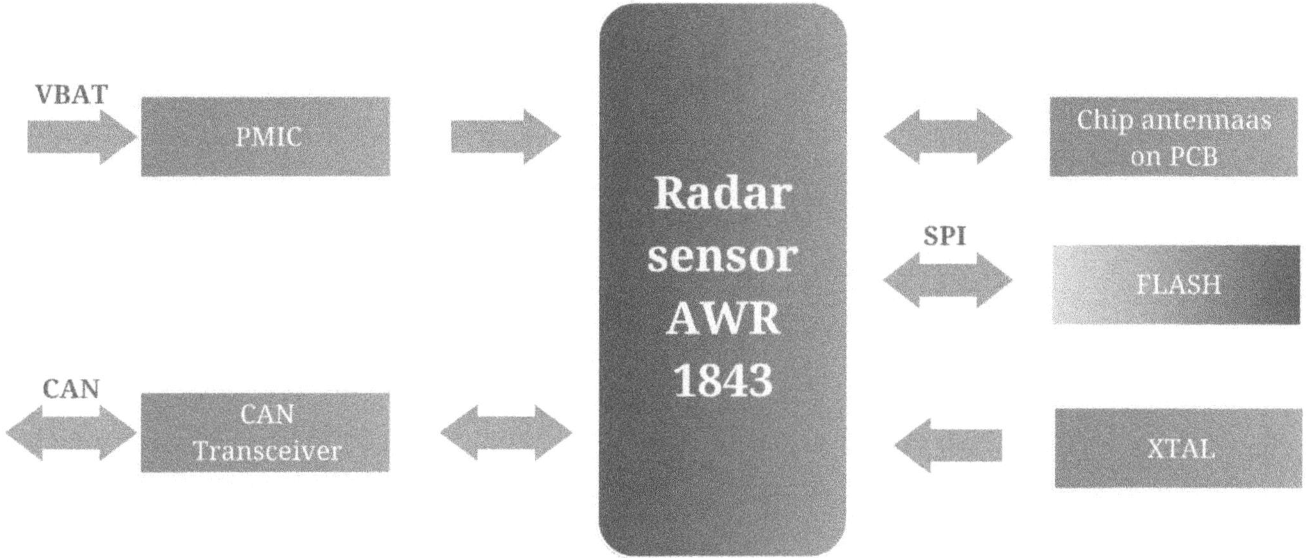

Here we take an example of AWR1843 Radar IC for sensing the rear-occupants in the cabin of the car. The ultrasonic sensor unit should have supporting components in the ECU mounted above the cabin (ROA sensor unit).

Examples of TSRs that can be defined for the above system:

Power Supply: In any mode, the 5V output from the PMIC shall be continuously monitored to be 4V< 5Voutput < 5.5V. Incase the 5V output is out of range, the AWR1843 shall be reset within FTTI1.

Microcontroller: An external watchdog shall be implemented for AWR1843 to reset the MCU if trigger is not received for FTTI1. The ROM memory shall be protected with ECC. Each time the data is read from ROM memory, the ECC correction shall be performed.

At each start-up, core self-test shall be performed.

At each start-up and at run time, readback shall be performed for configuration registers for every FTTI1.

At each start-up, the internal RADAR unit of AWR1843 shall be self-tested.

SPI interface: In normal mode, the CRC mechanism shall be implemented for all the safety-relevant data read from the SPI interface. Incase of wrong CRC detected, the SPI frame shall be discarded by AWR1843.

CAN interface: In normal mode, periodic CAN messages shall be sent every FTTI1 to establish continuous communication with the BCM module. In normal mode, CRC mechanism shall be implemented for the rear-occupant status data sent out by ROA sensor.

Bottom-up requirements: In normal mode, the timeout mechanism shall be implemented by BCM for the CAN interface for FTTI1. If timeout is detected, the BCM shall indicate to the driver that the ROA module is not alive.

FFI requirements: Freedom from interference shall be ensured by SW safety process so that any process in the QM function shall not affect the ASIL A function performed by the ROA module.

SEooC – System Element out of Context

Topics to be covered

1. What is the SEooC?

2. Examples of Assumptions made when developing an SEooC

3. The Bottom-up Approach to be applied for SEooC

4. Examples of SEooC Elements

5. Difference between In-context and Out of context System Element

6. Case Study

System Element Out Of Context (SEooC)

"System Element Out Of Context" (SEooC) is a term used in the context of functional safety standards, such as ISO 26262 for the automotive industry or IEC 61508 for general industrial applications.

In safety-critical systems, SEooC refers to a component or element that is designed, developed, and qualified for use in multiple systems without knowledge of the specific context in which it will be applied. This means that the component is developed to meet certain safety requirements and standards, but the final safety assessment and certification are performed within the context of the entire system in which it will be integrated.

The purpose of SEooC is to provide a more cost-effective approach to safety certification. Instead of re-evaluating and re-certifying every single component for each different application, SEooC allows the reuse of pre-qualified components in multiple systems as long as they are used within their specified limitations and conditions.

To ensure proper usage, SEooC components typically come with detailed documentation, guidelines, and restrictions for their integration into various systems. This documentation is essential for the system integrator to understand how to use the component safely and to ensure that the overall system still meets the required safety standards.

Keep in mind that safety standards and terminologies might evolve over time, so it's always a good idea to refer to the latest versions of the relevant standards or consult with experts in the field to get the most up-to-date information on SEooC or any related concepts.

The automotive industry develops generic elements for different applications and customers. These generic elements can be developed independently by different organizations. In such cases assumptions are made about the requirements and the design; including the safety requirements that are allocated to the element by higher design levels and on the design external to the element. Such elements can be developed by treating these as Safety Elements out of Context (SEooC).

As per definition from ISO26262 Part 10 [1], an SEooC is a safety-related element which is not developed for a specific item. This means, it is not developed in the context of a particular vehicle.

An SEooC can be a system, a combination of systems, a subsystem, a software component, a hardware component or a part. Examples of SEooC include system controllers, ECUs, microcontrollers, software implementing a communication protocol or an AUTOSAR software component.

An SEooC is thus developed based on assumptions; on an intended functionality and use context which includes external interfaces. These assumptions are set up in a way that addresses a superset of items, so that the SEooC can be used later in multiple different, but similar, items.

The validity of these assumptions is established in the context of the actual item while integrating the SEooC. An item may contain multiple SEooCs, with SEooCs interfacing directly to each other. In this case the validity of the assumptions of one SEooC is established considering the interfacing SEooC.

Examples of Assumptions made when designing SEooC:

Lets take an example of designing a Microcontroller Unit as SEooC, below would be the assumption made:

- Contribution of the MCU in violation of a safety goal

- Probability of failure of CPU instruction memory (derived using FMEDA)

- Safe State implementation by the MCU in case of a reset.

- Safety mechanisms implemented within the stipulated time

- ASIL value based on the assumed functionalities of the MCU

Similarly taking an example of developing a Software Unit as SEooC, the assumptions would be,

The software architecture of which the software will be a part of Interference caused by the software component is handled by the system. Assumption of ASIL value of the software based on the data it handles List of error conditions Handling of instances of Data corruption.

The Bottom-up Approach to be applied for SEooC

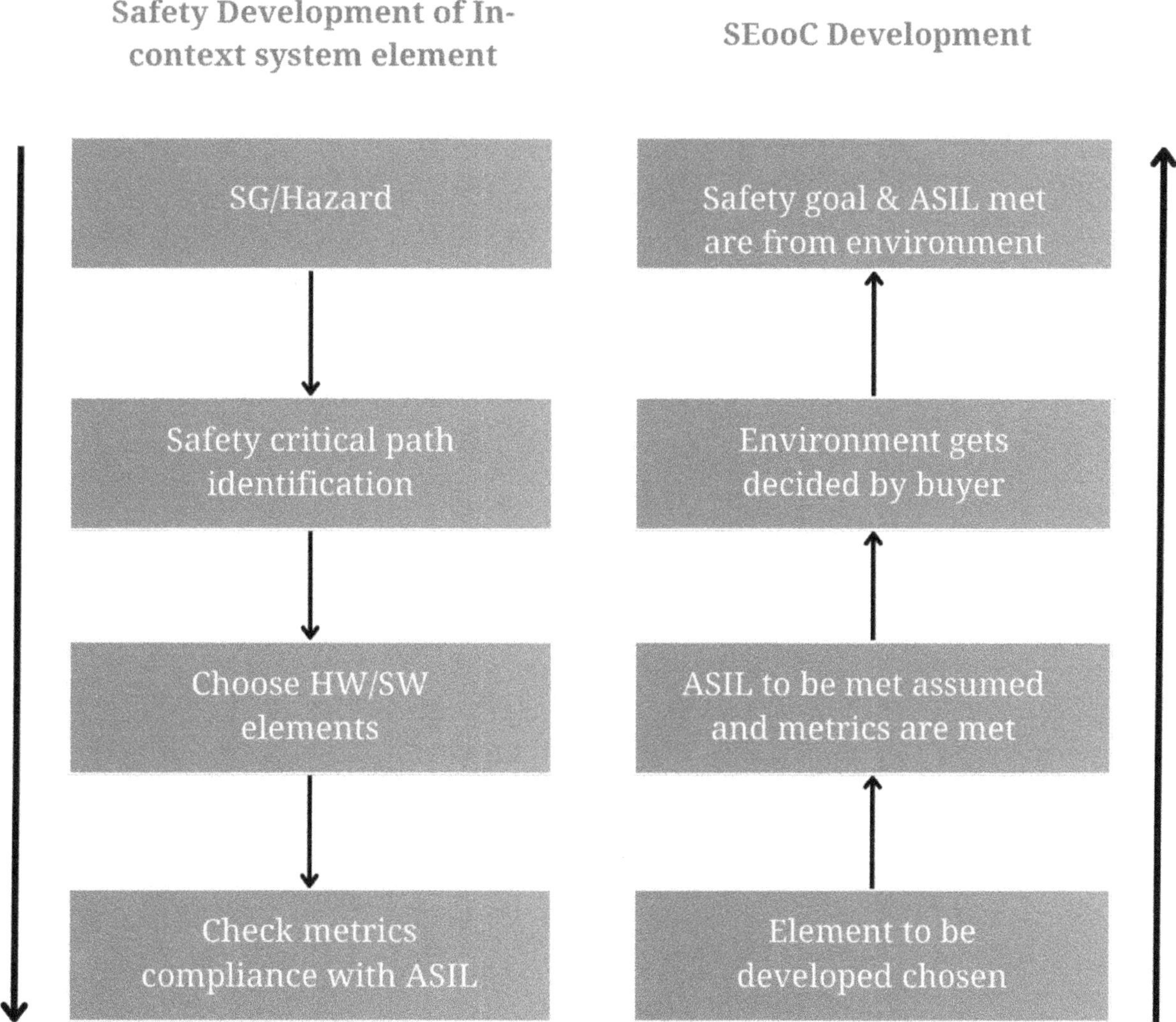

Safety Development of In-context System Element – Top-Down Approach

First, Hazards and associated Safety goals for an item for a specific vehicle are identified. Then the Safety path in the system for implementing that Safety goal is identified, then the Safety related HW, SW and System elements are decided, and finally these elements are developed in compliance with ASIL.

Safety Development of System Element out of Context – Bottom-up Approach

First, what is the HW, SW or System element that must be developed as ASIL is decide and then assumptions on the ASIL level, the Safety goals, the item or System, and the context/ environment in which the Safety element will be used are formulated.

Difference between SEooC and In-context System Element

The difference between System Element Out of context and and System element developed in context can be summarized by below table:

	SEooC	Safety element developed in context
Safety Goals	Assumed	Provided by OEM
Context of use (Vehicle, Operating conditions)	Assumed	Known/Pre-defined
Functional Safety requirements	Assumed for System SEooC Not Mandatory for HW and SW SEooC	Provided by OEM
Technical Safety Concept/Requirements	Assumed requirements defined or Requirements derived from Assumed FSR	Derived from FSR
Development process	Follow processes and methods in respective ISO Parts SW SEooC – Part 6 HW SEooC – Part 5 System SEooC – Part 4	Follow processes and methods in respective ISO Parts SW Safety Element – Part 6 HW Safety Element – Part 5 System Safety Element – Part 4
Safety Manual	Safety manual is mandatory to provide a set of "Assumptions of use" aka "external requirements" for the integrating item/system regarding how the SEooC must be used.	Not provided
Qualification	Audited/Assessed as SEooC. However, cannot be qualified standalone. Actual qualification of the SEooC happens in the Item only when all the assumptions of use are validated and satisfied by the Item	Audited/Assessed and Completely qualified in the Item context

Case Study

Let us take an example of a Lane Keep Assistance System, designed to warn the driver in case of lane departures.

This feature is supposed to be automatically activated at certain conditions. And can also be deactivated by the driver.

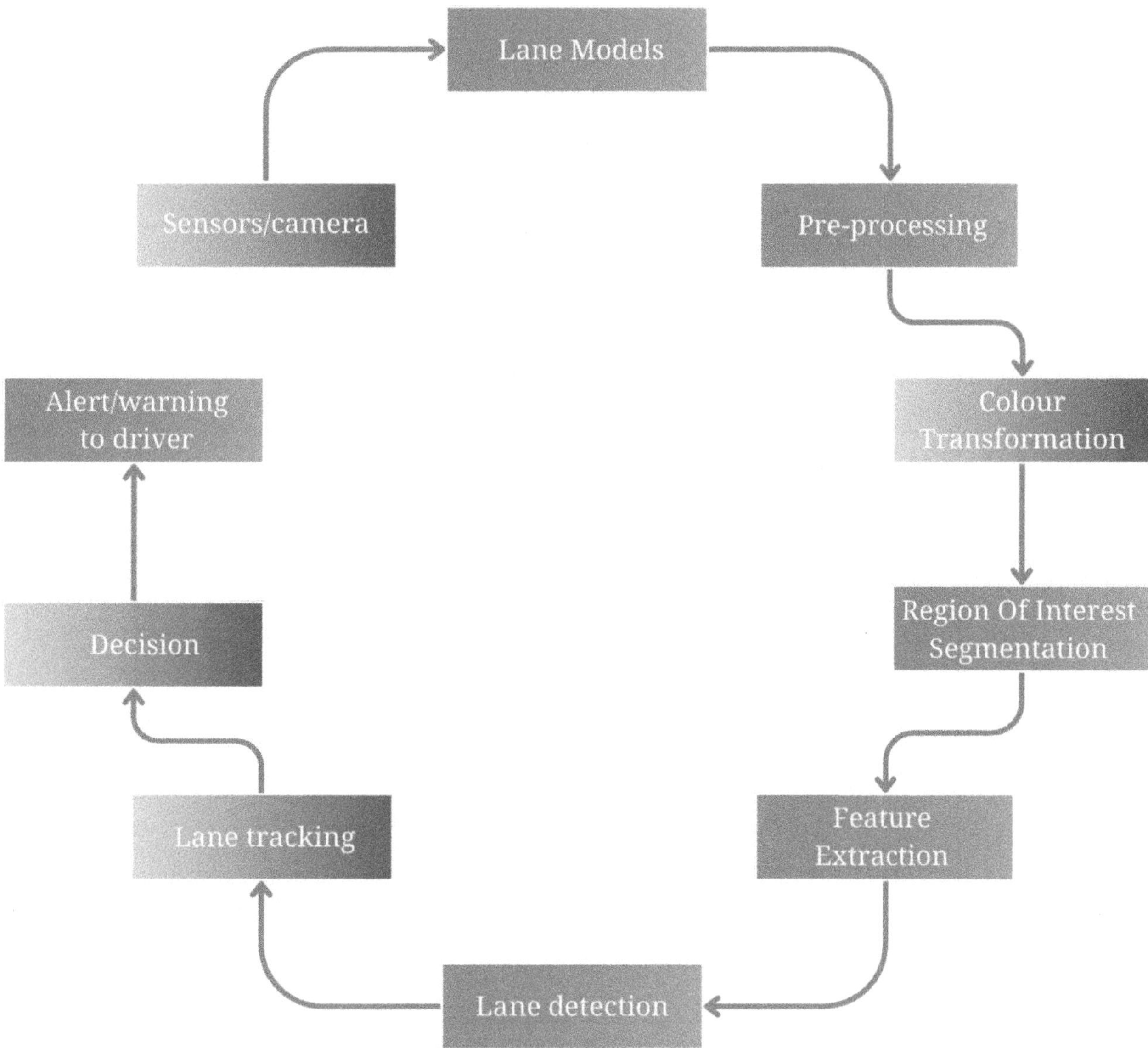

Examples of Function-based Assumptions made by the Developer

The system interfaces with other automotive ECUs via CAN BUS, in order to get the required vehicle info

It is designed for a front-wheel driven automotive

The system activates automatically as well as on driver's request- Functional requirement

The system deactivates if requested by the driver- Functional Requirement

ISO 26262 Concept-Phase Related Assumptions:

Item Definition to assume the components which with the system will interact

Safety Goals and ASIL Value

Functional Safety Requirements (FSR) and Technical Safety Requirement (TSR)

Example of Safety Goal and ASIL assumptions:

System shall not get activated at high-speed. - ASIL B

System shall not get deactivated unless the driver requests for it - ASIL B

The system shall fetch correct information from external ECUs about the vehicle speed and the driver's request - ASIL C

Chapter 6

ISO26262 Part 9 – ASIL decomposition

Topics to be covered

1. What is ASIL decomposition?

2. What is the need for ASIL decomposition?

3. Definition by ISO26262 Part 9

4. How it is done? The ASIL decomposition scheme by ISO26262

5. Inputs required for ASIL decomposition

6. Proof of Independence

7. Safety workproducts required for ASIL decomposition

8. Technical Safety concept - Safety requirements and Architectural information

9. Dependent Failure Analysis (DFA)

10. Case study

ASIL decomposition

Functional safety is indeed becoming a more mainstream requirement in the automotive industry. With the increasing complexity and integration of electronic systems in modern vehicles, ensuring their safe and reliable operation is of paramount importance to protect the vehicle occupants and other road users. Autonomous Emergency Braking (AEB) System, Electric power steering and airbag deployment are some of the common examples of ASIL D-rated applications in the automotive space. As automobiles continue to evolve and incorporate complex electrical/ electronic/programmable electronic (E/E/PE) dominated systems, the number of applications that require higher performance levels and the highest levels of diagnostic coverage continue to expand.

ISO 26262 ASIL Decomposition is defined in Part 9 of the ISO 26262 standard. Part 9 is titled "ASIL Decomposition and Assignment to Hardware and Software Elements." It provides guidelines and requirements for decomposing the overall Automotive Safety Integrity Level (ASIL) of the system into lower-level ASILs for individual hardware and software elements within the system. Part 9 of ISO26262 defines a scheme for dividing a requirement with a specific ASIL level into two requirements with lower ASIL levels. Higher ASIL rated applications have the potential to benefit from an ASIL-decomposed architecture that gives end-equipment designers increased flexibility when developing safety-critical systems with high ASIL requirements.

Need for ASIL decomposition

An Automotive Safety Integrity Level (ASIL)-decomposed architecture offers a reliable and robust path to achieving the highest levels of diagnostic coverage and gives end-equipment designers increased flexibility when developing safety-critical systems with high ASIL or SIL requirements.

ASIL decomposition is a method described in the ISO 26262 standard for the assignment of ASILs to redundant systems and its requirements. ASIL decomposition is not intended to reduce ASIL assignments to hardware elements for random hardware failures, but instead focuses on functions and requirements in the context of systematic failures.

Two potential reasons that may result in the use of "modified" ASIL algebra include the need of OEMs to partition a system and specify subsystem requirements to suppliers and the need for designers to construct systems bottom up.

Definition of ASIL decomposition by Part 9 ISO26262

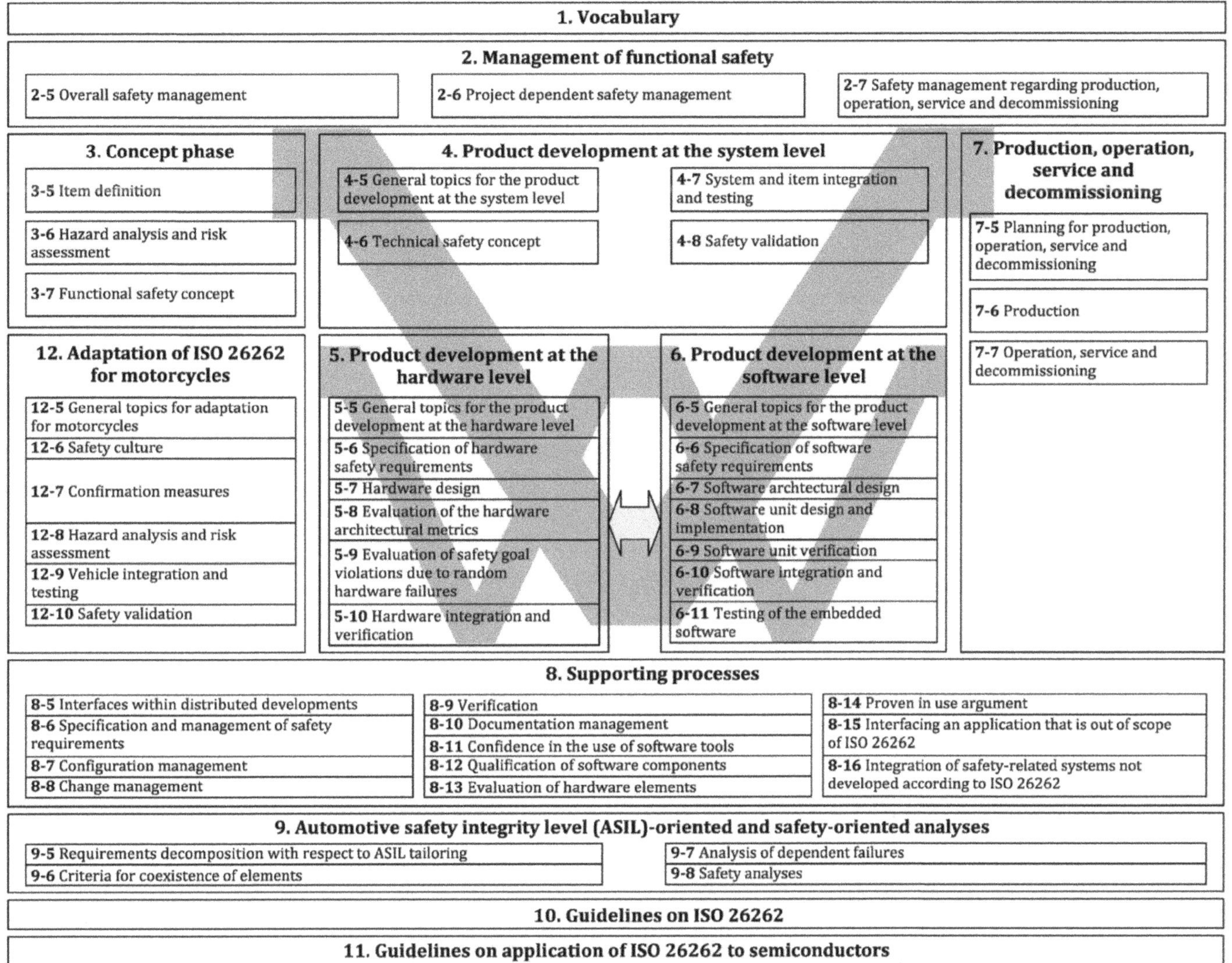

According to ISO26262 Part 9[1], below is the definition of ASIL decomposition:

"ASIL decomposition allows the apportioning of the ASIL of a safety requirement between several elements that ensure compliance with the same safety requirement addressing the same safety goal. "

The ASIL of the safety goals of an item under development is propagated throughout the item's development. Starting from safety goals, the safety requirements are derived and refined during the development phases. The ASIL, as an attribute of the safety goal, is inherited by each subsequent safety requirement. The safety requirements are allocated to architectural elements, starting with functional safety requirements allocated to elements of system architectural design and finally resulting in safety requirements allocated to the hardware and/or software elements.

ASIL decomposition is a method of ASIL customising during the concept and development phases. During the safety requirements allocation process, benefit can be obtained from architectural decisions including the existence of sufficiently independent architectural elements. This offers the opportunity

— to implement safety requirements redundantly by these independent architectural elements, and

— to assign a potentially lower ASIL to (some of) these decomposed safety requirements.

If the architectural elements are not independent enough, then the redundant requirements and the architectural elements inherit the initial ASIL.

ASIL decomposition is an ASIL tailoring measure that can be applied to the functional, technical, hardware or software safety requirements of the item or element.

ASIL decomposition schemes by ISO26262

ISO26262 Part9 defines a set of decomposition schemes that can be implemented while partitioning a higher ASIL system to sub-systems with lower ASILs.

Below image provides various schemes defined by the Standard to split an Element E with higher ASIL to two different subsystems E1 and E2 with lower ASILs and the requirements defined by the clause 5.4.9 in standard which are to be satisfied to achieve the decomposition.

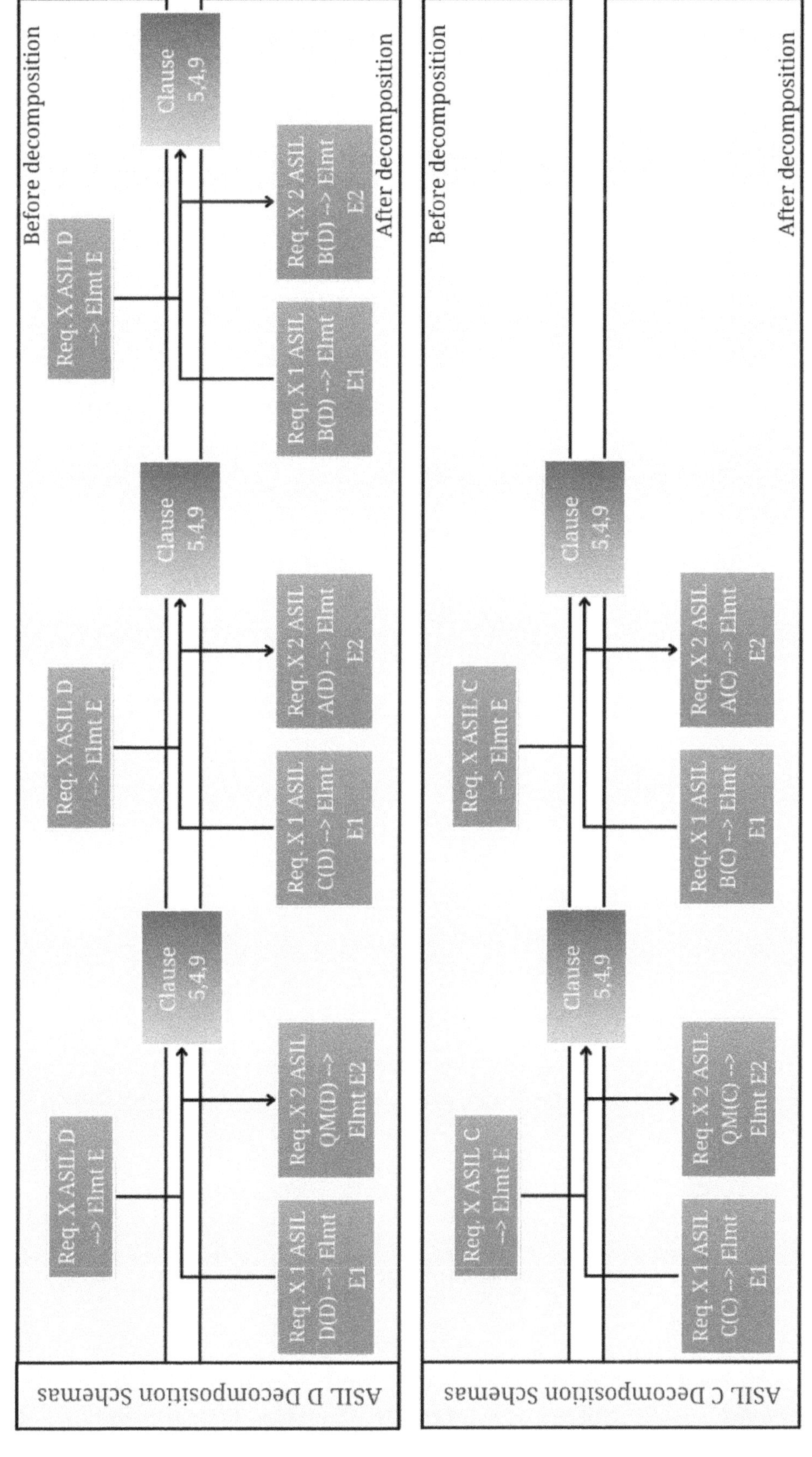

Before decomposition
After decomposition
ASIL D Decomposition Schemas
Clause 5.4.9
Clause 5.4.9
Clause 5.4.9
Req. X ASIL D --> Elmt E
Req. X ASIL D --> Elmt E
Req. X ASIL D --> Elmt E
Req. X 2 ASIL B(D) --> Elmt E2
Req. X 1 ASIL B(D) --> Elmt E1
Req. X 2 ASIL A(D) --> Elmt E2
Req. X 1 ASIL C(D) --> Elmt E1
Req. X 2 ASIL QM(D) --> Elmt E2
Req. X 1 ASIL D(D) --> Elmt E1
Before decomposition
After decomposition
ASIL C Decomposition Schemas
Clause 5.4.9
Clause 5.4.9
Req. X ASIL C --> Elmt E
Req. X ASIL C --> Elmt E
Req. X 2 ASIL A(C) --> Elmt E2
Req. X 1 ASIL B(C) --> Elmt E1
Req. X 2 ASIL QM(C) --> Elmt E2
Req. X 1 ASIL C(C) --> Elmt E1

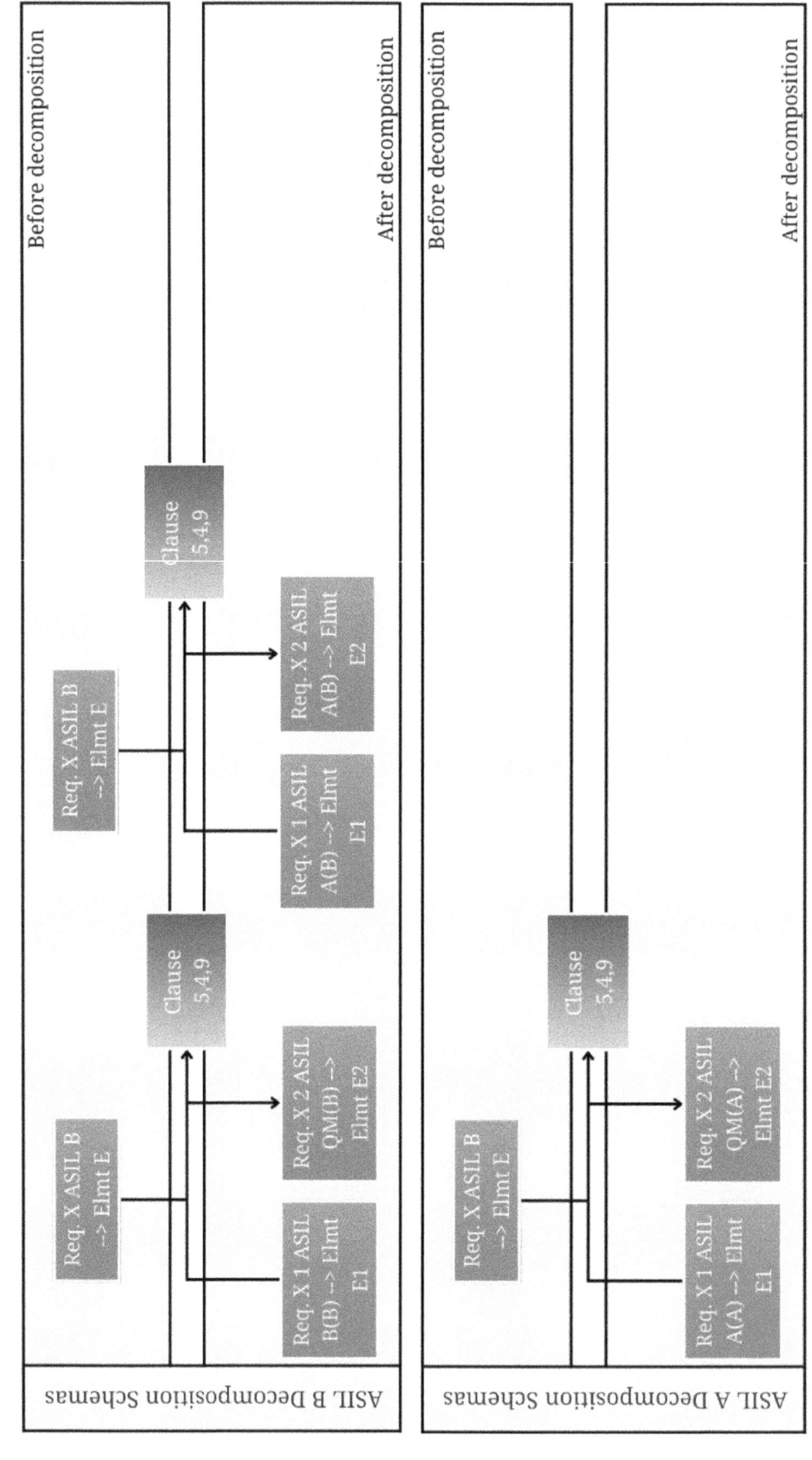
Before decomposition
After decomposition
Before decomposition
After decomposition
Clause 5,4,9
Clause 5,4,9
Clause 5,4,9
Req. X ASIL B -> Elmt E
Req. X 1 ASIL A(B) -> Elmt E1
Req. X 2 ASIL A(B) -> Elmt E2
Req. X ASIL B -> Elmt E
Req. X 1 ASIL B(B) -> Elmt E1
Req. X 2 ASIL QM(B) -> Elmt E2
Req. X ASIL B -> Elmt E
Req. X 1 ASIL A(A) -> Elmt E1
Req. X 2 ASIL QM(A) -> Elmt E2
ASIL B Decomposition Schemas
ASIL A Decomposition Schemas

The schemes that can be implemented as per the above image are:

1. An ASIL D requirement shall be decomposed as one of the following:

    a. one ASIL C(D) requirement and one ASIL A(D) requirement; or

    b. one ASIL B(D) requirement and one ASIL B(D) requirement; or

    c. one ASIL D(D) requirement and one QM(D) requirement.

2. An ASIL C requirement shall be decomposed as one of the following:

    a. one ASIL B(C) requirement and one ASIL A(C) requirement; or

    b. one ASIL C(C) requirement and one QM(C) requirement.

3. An ASIL B requirement shall be decomposed as one of the following:

    a. one ASIL A(B) requirement and one ASIL A(B) requirement; or

    b. one ASIL B(B) requirement and one QM(B) requirement.

4. An ASIL A shall only be decomposed, if needed, as one ASIL A(A) requirement and one QM(A) requirement.

Inputs required for ASIL decomposition

As per ISO26262 Part 9, the prerequisites to perform ASIL decomposition are:

*   item definition from Concept phase; and

*   safety goals derived in the hazard analysis and risk assessment report

*   the safety requirements at the level at which the ASIL decomposition is to be applied: vehicle, system, hardware, or software

*   the architectural information at the level at which the ASIL decomposition is to be applied: vehicle, system, hardware, or software

Co-existence of Elements: Proof of Independence

By default, when an element is composed of several sub-elements, each of those sub-elements is developed in accordance with the measures corresponding to the highest ASIL applicable to the element, i.e. the highest ASIL of the safety requirements allocated to the element.

In the case of the coexistence of sub-elements that have different ASILs assigned, or the coexistence of non-safety-related sub-elements with safety-related ones, it can be beneficial to avoid assigning the ASIL of the element to all the sub-elements. For this purpose, the standard provides guidance for determining if sub-elements with different ASILs can coexist within the same element. This has to be performed based on the analysis of interference of each sub-element with the other sub-elements of an element.

For example the ASIL D Decomposition scheme is implemented in the target system, then Proof Of Independence or Freedom From Interference can be proved as explained:

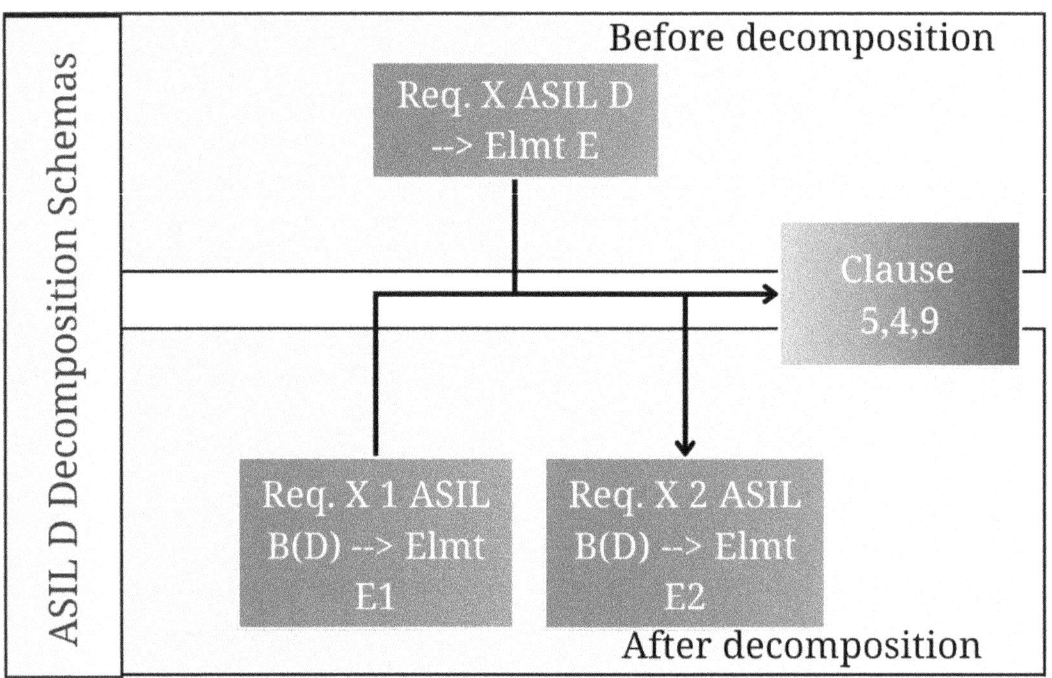

In the above system, the independence between Element E1 and Element E2 must be proved and validated. Independence directly means the ability of one element to work and control the Safety goal even if the other element completely fails. This shall be confirmed by FSC, TSC and DFA.

There are two failures that have to be proved to be protected by Safety mechanisms to achieve freedom from independence between the partitioned elements.

Common-cause failure

Cascading failure

The interference is the presence of cascading and common cause failure that can happen in a sub-element with a lower ASIL assigned and can penetrate to a sub-element with a higher ASIL assigned which leads to the violation of a safety requirement of the element.

Common cause failure

Common cause failure is when a single failure affects both the ASIL rated component used under the ASIL decomposition scheme.

Common Verification when choosing a power supply for the target system would be to check Current consumption, Regulated output, Figure of merit.

When ASIL decomposition is involved, even if SBC is capable of regulating the current requirement of both MCUs, a common power supply means when the power supply fails (undervoltage, overvoltage, no output), then both the MCUs will enter reset, leaving the system unable to reach Safe state

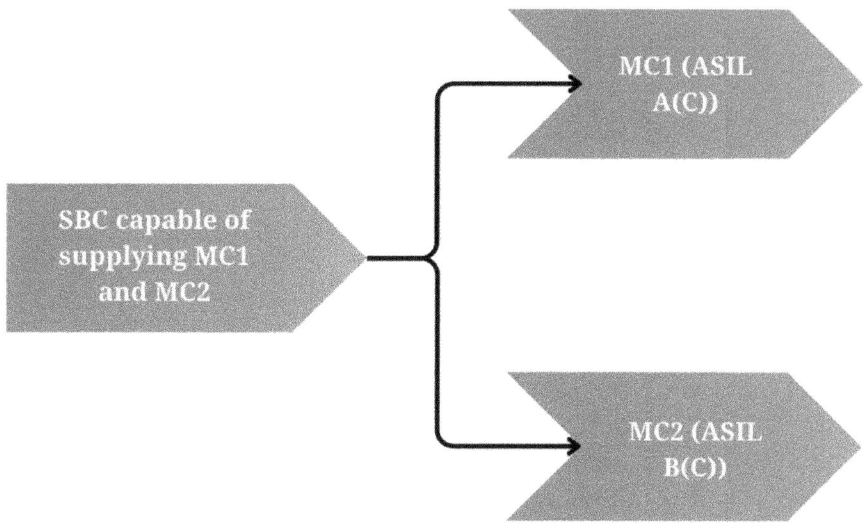

The above image shows a common cause failure, where if the SBC fails both MC1 and MC2 will together fail. Possible countermeasure for the above scenario would be to use Redundant power supplies

Cascading failure

Cascading failure is similar to a chain reaction which finally affects the Safety goal.

Eg: Program memory corruption. If the corruption is memory leading to failure to call and wake up error monitoring part of the SW code. This can boot the MCU and in normal condition function properly, but when error occurs, it fails to detect and reach safe state with the fault tolerant time, hence violating the Safety goal.

Note: All the counter-measures that are identified shall be defined in the Safety concept, so that the same will be followed during the implementation of Hardware and Software elements in the upcoming phases of V-cycle. In addition to that, the architecture has to be validated to confirm the elimination of the failure, which will be done across the respective safety requirements defined in the safety concept.

Safety work products required for ASIL decomposition

The ASIL decomposition requires Safety work products to confirm Proof of Independence

- Functional Safety Concept - Sub-system level proof of independence is defined

- Technical Safety Concept - Architecture level proof of independence is defined

- Dependent Failure Analysis - The resultant architecture from TSC implementation is analysed to have any dependencies on one another components.

Functional/Technical Safety Concept - Safety requirements and Architectural information

In Safety Concept, below requirements shall be clearly stated

- Safety Requirements stating the proposed ASIL decomposition scheme.

- Safety Requirement requesting Freedom from interference.

In case of any common cause or cascading failure is expected according to the architecture, then test requirements shall be defined to validate independent working of Hardware and Software components. The test methods to be covered shall be clearly stated.

Examples:

TSR_1: The microcontroller's MC1 and MC2 shall be implemented with different core architectures. (ASIL C)

TSR_2: The Freedom from Interference shall be confirmed by Software Critical Path Analysis (ASIL C)

TSR_3: The memory unit used by the ASIL B rated HAF status evaluation shall not be interfered by QM or ASIL A rated functions (ASIL B)

During Safety concept definition, below is the process to follow to implement the ASIL decomposition:

- Freeze the basic architecture

- Identify safety critical signals

- Check for possible counter measures to achieve ASIL without decomposition

- Check for counter measure by applying ASIL decomposition

- Check whether the counter measure is desired ASIL qualified (must for the alternate path hardware and software to be ASIL rated depending on the ASIL decomposition scheme)

- In case the ASIL metrics are not met, apply necessary Safety mechanisms and perform FMEDA to qualify the part

- Confirm Independence between the Main path and the counter measure (FSC or TSC)

- Apply Freedom from Interference Safety mechanisms to confirm the above (FSC or TSC)

- In case of potential CCF or CF identified, define test requirements to prove the independence (DFA)

Dependant Failure Analysis (DFA)

The scope of the analysis of dependent failures can be influenced by the technology of the given elements (e.g. software elements, hardware elements, or a mix of hardware and software elements), and by the safety requirements involved.

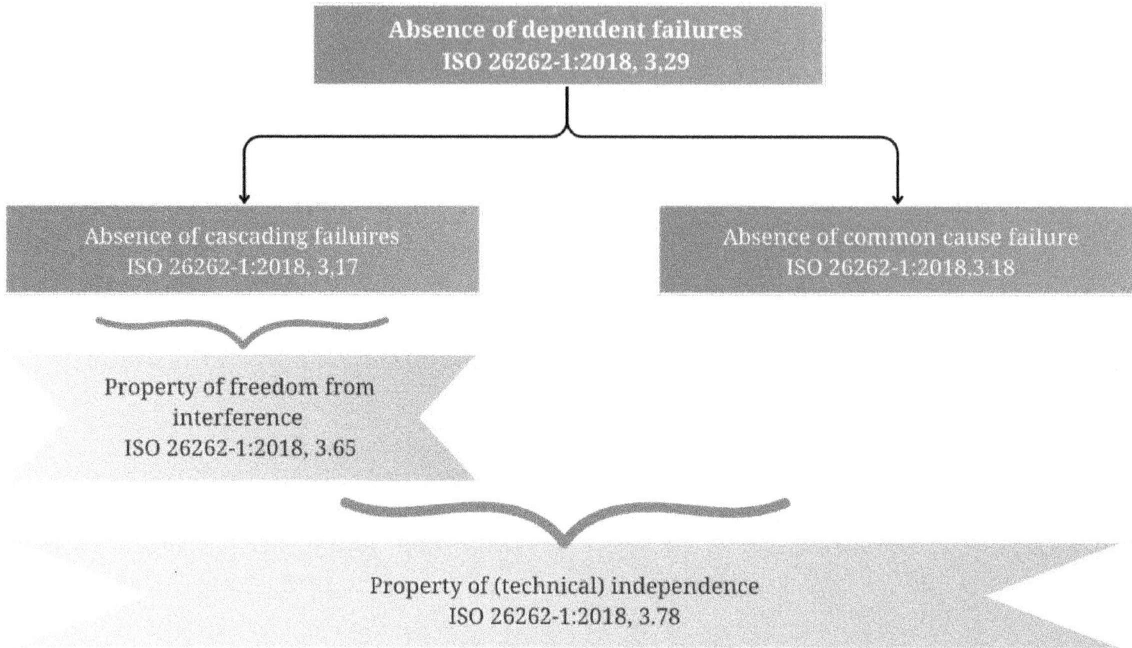

The work product DFA- Dependant Failure Analysis is performed at design validation stage to confirm Freedom from interference.

DFA identifies Common Cause failure and Cascading failures that can jeopardize the system from detecting the failure and reaching the safe state, that are not solved by design countermeasures.

Below is the template that would be followed for the analysis:

TSR	ASIL	Elements involved for ASIL decomposition	Potential Root caused of CCF or CF	Existing design counter-measures	Likehood of CCF or CF after counter-measures?	Incase yes, Additional measures design or justification by tests, analyses, field returns...

The template analyses the following:

- FSR/TSR - the Safety requirement defined by the safety concept would be the rules under which the architecture will be analysed

- ASIL - ASIL to be satisfied by the architecture element for the corresponding safety requirement would be marked

- Potential Root cause for CCF or CF - Any potential root cause that can lead to a common cause or cascading failure of the architectural element shall be listed for analysis. Eg: Same power supply architecture for both sub-elements.

- Existing design measures - Any existing design measure that is defined by the Safety concept will be mentioned Eg: Different Power monitoring circuit for both sub-elements.

- Likelihood of CCF or CF after measure - this would be the analysis result where we finalise whether the countermeasure is actually eliminating the common cause or cascading failure identified.

- Incase yes, Additional Measures - If yes, then further analysis would be required as to whether we can confirm the independence by conducting tests or any further safety measures can be implemented to fully eliminate the failure.

1. The Standard defines to analyze the architecture across following requirements:

    - The evaluation of the plausibility of the potential dependent failures can be supported by appropriate checklists, e.g. checklists based on field experience. The checklists provide the analysts with representative examples of root causes and coupling factors such as same design, same process, same component, same interface, proximity. Annex C of ISO26262 Part 9 [1] can be used as a basis to establish such checklists.

2. This evaluation can also be supported by the adherence to process guidelines which are intended to prevent the introduction of root causes and coupling factors that could lead to dependent failures.

    a. random hardware failures;

3. Eg: Failures of common blocks such as clock, test logic and internal voltage regulators in large scale integrated circuits (microcontrollers, ASICs, etc.).

b. development faults;

4. Eg: Requirement faults, design faults, implementation faults, faults resulting from the use of new technologies and faults introduced when making modifications.

c. manufacturing faults;

5. Eg: Faults related to processes, procedures and training; faults in control plans and in monitoring special characteristics; faults related to software flashing and end-of-line programming.

d. installation faults;

6. Eg: Faults related to wiring harness routing; faults related to the interchangeability of parts; failures of adjacent items or elements.

e. service faults;

7. Eg: Faults related to processes, procedures and training; faults related to trouble-shooting; faults related to the interchangeability of parts and faults due to backward incompatibility.

f. environmental factors;

8. Eg: Temperature, vibration, pressure, humidity/condensation, pollution, corrosion, contamination, EMC.

g. failures of common external resources or information;

9. Eg: Power supply, input data, inter-system data bus and communication.

h. stress due to specific situations; and

10. Eg: High operational loads, extreme user inputs or requests from other systems, thermal impact, and mechanical shocks.

i. aging and wear.

Few recommended design requirements to confirm Independence

- Always use independent power supplies and watchdog units.

- Incase of processor application, use separate memories with independent Safety mechanisms.

- If design permits, use separate command and control interfaces.

- Use Microcontrollers with different core architecture.

- Best case, use microcontrollers from different manufacturers. (same manufacturers have similar EMC, ESD effects on HW components)

In case of unavoidable situations

In a case where OEM recommends using same MCU for cost saving purposes, the test requirements shall be defined using DFA.

Below is such an example:

TSR	ASIL	Elements involved for ASIL decomposition	Potential Root caused of CCF or CF	Existing design countermeasures	Likehood of CCF or CF after countermeasures?	Incase yes, Additional measures design or justification by tests, analyses, field returns...
The two independent MC1 and MC2 shall be dissimilar.	C	MC1 ASIL A(C) and MC2 ASIL B(C)	Environmental factor-- EMC/Electro-magnetic field	- Architecture independence between MC1 and MC2; - EMC protection in Power supply filter	YEX	- EMC test: Repetitive EMC test to meet OEM specification standards for atleast 10 test samples

In such cases, test cases and requirements shall be clearly defined that are to be followed during the Design and Product validation phase to confirm that the identified Root cause is not occurring in the product and will not be violating the Safety goal under real use case conditions.

Case study

Electronic Steering Lock [2]

A steering lock is a bolt driven into the steering column when the steering wheel is completely turned or when the vehicle is locked. The main purpose of the steering wheel lock is to prevent illegal operation and use as an anti-theft device. The functional safety goal of a steering lock is that it shall not engage unintentionally while the vehicle is being driven. The consequences of the steering lock engaging unintentionally could be catastrophic; therefore, this function is rated ASIL D.

The below figure shows a preliminary architecture where a body control module (BCM) communicates over the Controller Area Network (CAN) bus with the MCU, which in turn drives an actuator through a bridge driver to drive the bolt into the steering column and lock it.

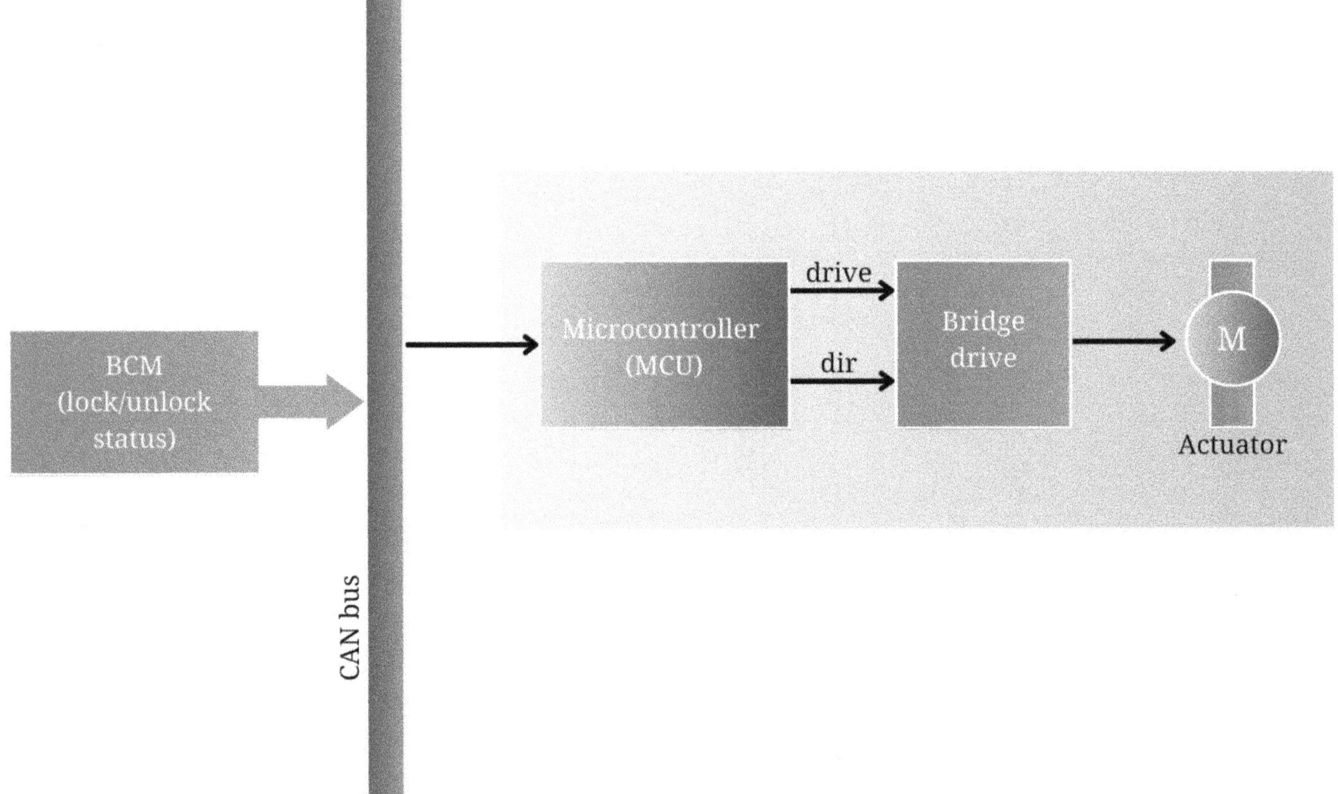

This is a single-channel architecture and requires an MCU to drive an H-bridge to meet the random hardware metrics of ASIL D. Design Constraints could force the use of an MCU that is not rated to ASIL D. Even if the MCU met the ASIL D metrics, a single MCU that communicates with the BCM could still have a communication failure that leads to engaging the steering lock while the vehicle is in motion. This can be addressed by decomposing this requirement.

To appreciate the correct way to use ASIL decomposition, let's briefly review an incorrectly decomposed implementation.

In the flawed architecture shown below, the BCM sends a command to either lock or unlock the steering column. There is an additional antilock braking system/electronic skid protection module (ABS/ESP) that communicates the vehicle's speed over CAN. The primary MCU is responsible for driving the actuator operating the bolt into the steering column. The secondary MCU is responsible for enabling the H-bridge only if the vehicle is at rest. However, the secondary MCU is receiving the vehicle speed information through the primary MCU; there is no independent connection to the vehicle's CAN bus.

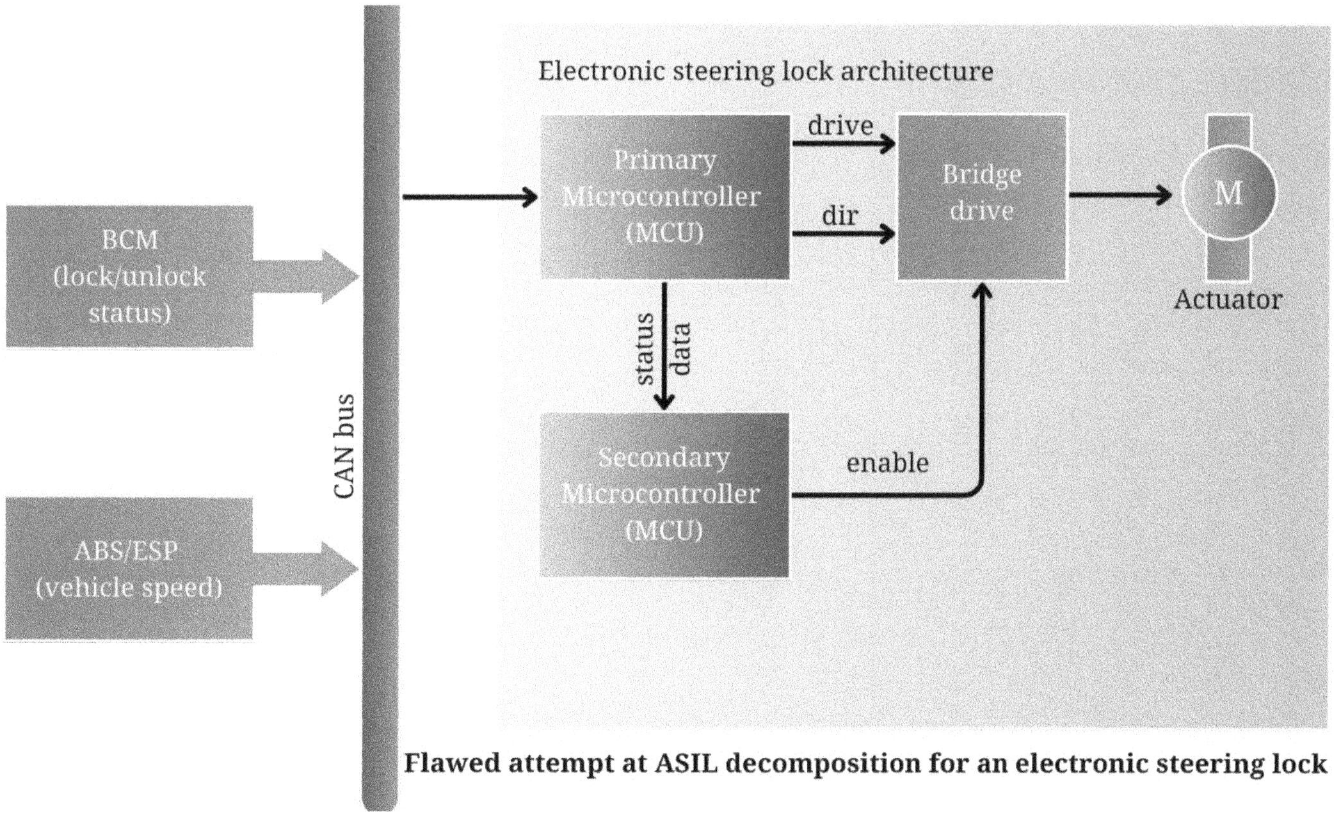

Flawed attempt at ASIL decomposition for an electronic steering lock

Consequently, a common cause failure on the primary MCU could result in incorrect vehicle speed information being transmitted to the secondary MCU. This could lead to the probability of violation of the system safety goal.

A better method of decomposing the steering lock example is as below:

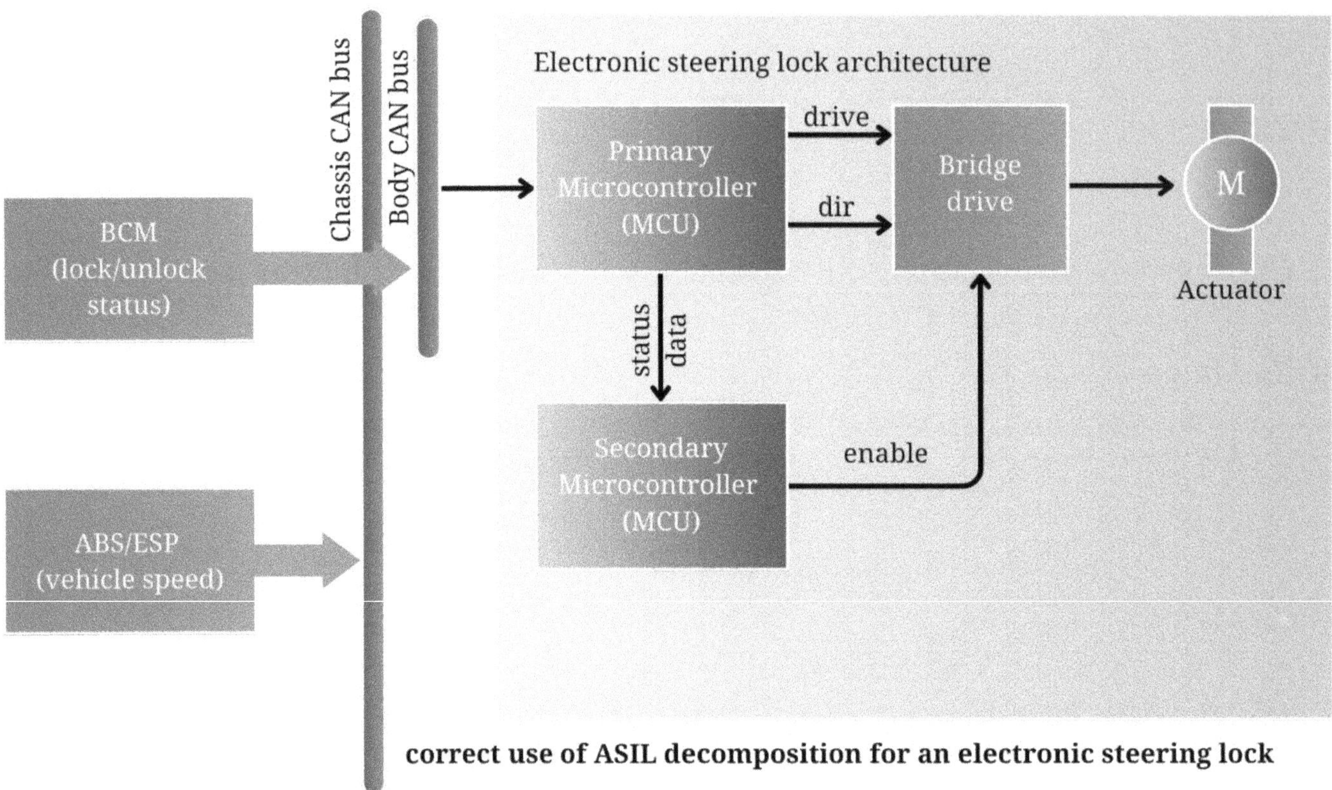

correct use of ASIL decomposition for an electronic steering lock

Using two independent CAN buses helps avoid common cause failures. The BCM module uses the body CAN bus to send lock/ unlock commands to the primary MCU. The ABS/ESP module uses the chassis CAN bus to communicate with the secondary MCU, which will enable the H-bridge only when the vehicle is stopped.

In addition to being a more robust way to achieve the safety goal, this ASIL-decomposed architecture also enables the use of two ASIL B MCUs (as governed by the ASIL decomposition rules in ISO 26262) to implement an ASIL D function.

Advantages of implementing ASIL decomposition

An ASIL-decomposed architecture offers a more reliable and robust path to achieve the highest levels of diagnostic coverage through these advantages:

- It's possible to use existing components like an ASIL B or ASIL A MCU to implement a higher ASIL D system.

- It's possible to deploy a complex or legacy software codebase on an ASIL QM MCU while programming an ASIL D safety MCU to implement lower-complexity supervisory functions. An example would be an electric vehicle (EV) traction inverter/motor control running on an ASIL QM MCU and a safety supervisor MCU that can de-energize the complex-control MCU in the event of a fault.

- By implementing a decomposed architecture as dual channels, there is the added possibility of achieving a true fail-operational capability that the system could rely on until the completion of appropriate repairs or system recovery. In contrast, an architecture implemented on a single device would only have the ability to de-energize as a safe-state response in the event of a catastrophic failure such as a loss of power.

- An added benefit of an ASIL-decomposed dual-channel architecture is the flexibility to choose MCUs based on inherent strengths so that system integrators do not have to sacrifice their system performance goals. For example, the acceleration performance of an EV traction motor or efficiency of a DC/DC converter does not have to be compromised by the limited pulse-width modulation (PWM) performance of a general-purpose ASIL D MCU.

Chapter 7

Part 4 – Hardware Software Interface

Topics to be covered

1. Hardware Software Interface Layer

2. Safety and HSI documentation

   a. Need of HSI documentation

   b. What should be covered in the HSI documentation?

   c. What are the mandatory components to be covered?

3. Safety critical HSI data on sample components

   a. HSI on SoC

   b. HSI on Switches

   c. HSI on Memory

   d. HSI on Sensors

   e. HSI on Power Supplies

Hardware Software Interface (HSI) Layer

"It's the layer at which the software "talks" to the hardware."

In the context of the modern-day CPU, the HSI and the Instruction Set Architecture (ISA) are essentially the exact same thing. It's the layer at which the software "talks" to the hardware. The CPU could be ARM, RISC-V - it doesn't actually matter because the process stays the same. The HSI Layer is the boundary between the hardware and software components of an electronic system. It defines how the software interacts with the underlying hardware to perform safety-critical functions. This layer is particularly important in systems that have safety requirements, as it helps manage the communication and control between hardware and software in a way that ensures the system's safe and reliable operation.

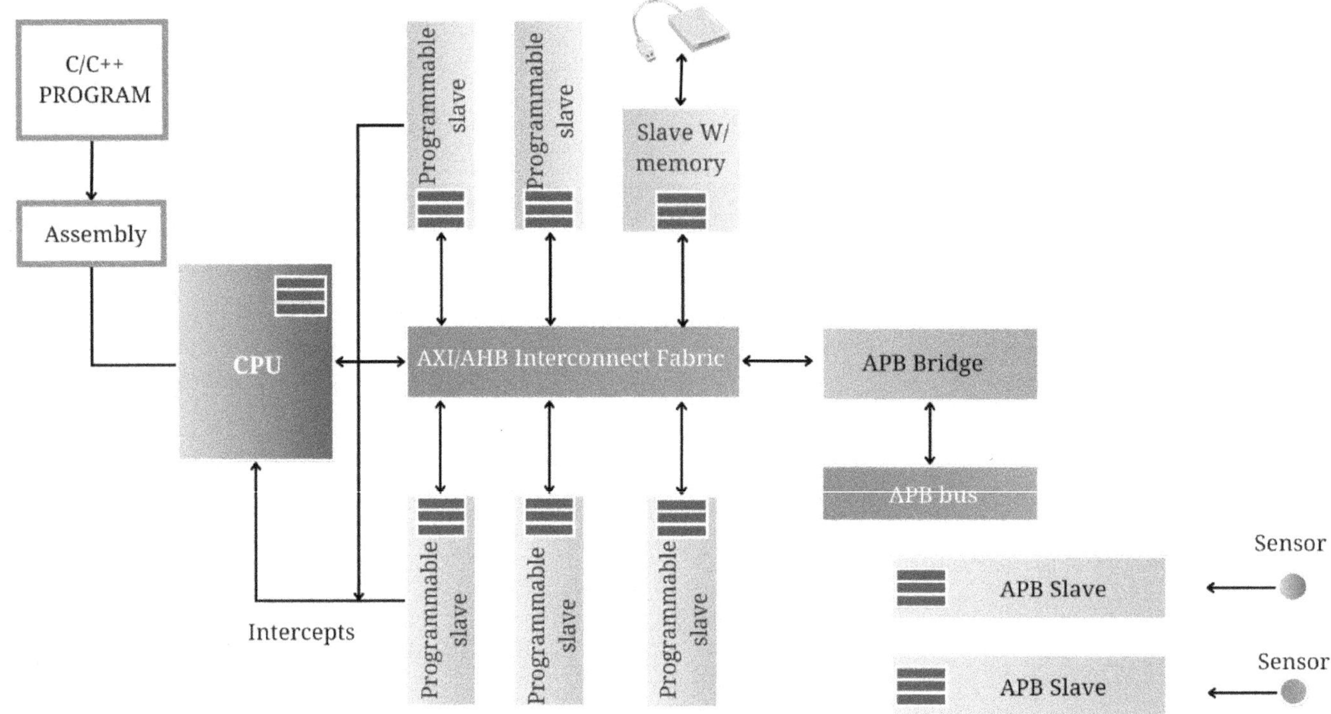

Regarding the actual SoC, you also have to deal with the network interface that connects the CPU to various dependent services. Depending on the specific situation you're talking about, these slaves may have their own memory or even a bridge to the delay bus. Slaves are programmed by reading and writing registers. When you look at things from this macro perspective, registers and interfaces are IP (or slave) HSI.

Safety and Hardware Software Interface (HSI) Documentation

1. The document describes both the configuration and the functionality of SoC and other IC peripherals and how they interact with CPUs

2. The Hardware is power supplies, SoC default configurations, hard-coded monitors, memories etc.

3. The Software is logic, flowchart, function calls, loop execution, etc.

4. The HSI takes care of

a. Resource allocation – Resource allocation is the process by which a computing system aims to meet the hardware requirements of an application run by it. Computing, networking and energy resources must be optimized taking into account hardware, performance and environmental restrictions.

b. I/O logics – Input-output (I/O) systems transfer information between computer main memory and the peripherals. An I/O system is composed of I/O devices (peripherals), I/O control units, and software to carry out the I/O transaction(s) through a sequence of I/O operations.

c. Peripheral allocations

d. IC to IC interfaces

e. Typical SoC Architecture

f. CPU & IP Hardware Software Interface

g. Register/Memory related

h. Types of Registers

i. Abstracting Register Buses

j. Read-Modify-Write/Mirror Memory Map

k. Register Sequences

l. Specification formats

m. Possible outputs

Need of HSI document

In modern electronic systems, devices and drivers are usually manufactured separately, because their development processes need highly different expertise. Driver development commonly requires the device's presence in order to exercise the functionalities and test the implementation conformance to Hardware/Software (HW/SW) interface protocols

- The sheer volume of different factors here - from register bits to access types, properties and the functionality they control - can be absolutely staggering in a modern SoC.

- HSI is required as a design specification document for the following purposes

- To define the layer of HW that takes directly with the SW and vice versa

- To define clearly the translation between voltage values and flowchart values

- If HSI layer is defined, major number of design flaws can be eliminated.

What should be covered by HSI?

- SOC configuration – The main components of an SoC typically include a central processing unit, memory, input and output ports, peripheral interfaces and secondary storage devices. The core of the SOC are configurable and the configuration data,

- Multi core functionality description – System-on-chip contains several

- Boot configuration

- ADC functionality

- Switches control and functionality

- Memory unit configurations

- Sensor configuration and parameters

Components to be covered in the HSI

All the major ICs and the supporting ICs that can directly impact any Safety goal of the system are to be mandatorily covered in the document.

Some of the examples are

- SOC

- PMIC

- DDR/Memories

- Watchdog

- Switches

- Serializer /deserializer

Safety critical HSI data on components

HSI on System On chip(SoC)

Automotive system-on-chip refers to a microchip installed in automotive devices which provides high system integration and less power consumption. The automotive system-on-chip helps manufacturers to increase safety and convenience experience level for the driver's-seat.

Automotive System-On-Chip is used to integrate functions in a car including sensing, controlling and actuating a motor. This helps to save space and energy, improve overall system reliability with diagnostic features and reduces overall cost owing to minimizing the number of components.

This SoC employed in automotive enables safety systems and driver assistance systems. Some of the intelligent and safety-critical functions enabled are blind-spot detection systems, backup cameras, collision-avoidance sensors, adaptive cruise controls, lane-change assist, airbag deployment sensors and emergency braking systems.

Few mandatory specifications to be defined for the SOC are:

- GPIO configuration

- Clock Configuration

- Power supply configuration

- Peripheral configuration

- Reset configuration

- Memory configuration

Case Study of HSI on SoC

Taking below Example of SoC architecture

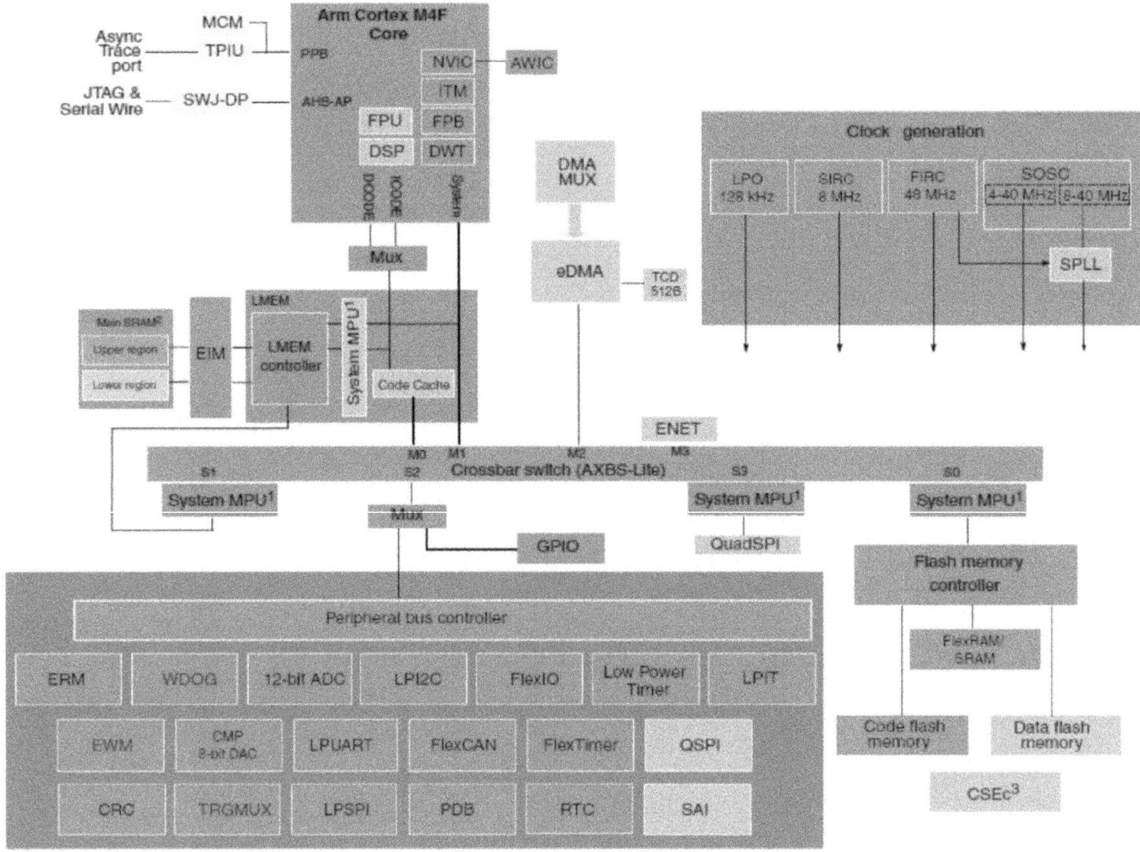

Functionality of a SoC will be mainly based on

- Register configuration

- Clock configuration

- Peripheral enable/disable

- Peripheral multiplexing

- Data rate for peripherals and corresponding registers to be configured

- Internal self-test blocks enable/disable

- Power control for each internal block

- Clock control for each internal block

- Default Pull up/pull down/open drain configuration

- Power pins and power supply tolerance ranges

- Clock pins and tolerance ranges

- Boot configuration and boot memory peripheral (incase external)

- ADC inputs and their ranges.

- The binary value to be noted for each ADC input range

GPIO configuration to be implemented for the SoC

#	Peripheral	Signal	Power group	Direction	Drive(#3 bit)	Pull Select(#28FDSOI)	wakeup control	output/input configuration
AP26	MIPI_CSI0_I2C0	i2c_scl	VDD_MIPI_CSI_DIG_1P8 (1.8V)	Input/Output	0b111: High_Speed	0b11: No pull	0b000: OFF	0b00: DEFAULT
AM24	MIPI_CSI0_I2C0	i2c_sda	VDD_MIPI_CSI_DIG_1P8 (1.8V)	Input/Output	0b111: High_Speed	0b11: No pull	0b000: OFF	0b00: DEFAULT
U35	ADMA_ADC	adc_in, 0	VDD_ADC_DIG_1P8 (1.8V)	Input	No init	0b11: No pull	0b000: OFF	0b11: INOUT
U33	ADMA_ADC	adc_in, 1	VDD_ADC_DIG_1P8 (1.8V)	Input	No init	0b11: No pull	0b000: OFF	0b11: INOUT
V32	ADMA_ADC	adc_in, 2	VDD_ADC_DIG_1P8 (1.8V)	Input	No init	0b11: No pull	0b000: OFF	0b11: INOUT
V30	ADMA_ADC	adc_in, 3	VDD_ADC_DIG_1P8 (1.8V)	Input	No init	0b11: No pull	0b000: OFF	0b11: INOUT
W29	ADMA_ADC	adc_in, 4	VDD_ADC_DIG_1P8 (1.8V)	Input	No init	0b11: No pull	0b000: OFF	0b11: INOUT
V34	ADMA_ADC	adc_in, 5	VDD_ADC_DIG_1P8 (1.8V)	Input	No init	0b11: No pull	0b000: OFF	0b11: INOUT
AJ31	SCU_DSC	dsc_boot_mode, 0	VDD_ANA1_1P8 (1.8V)	Input	No init	0b10: Pull-down	0b000: OFF	0b11: INOUT
AK32	SCU_DSC	dsc_boot_mode, 1	VDD_ANA1_1P8 (1.8V)	Input	No init	0b10: Pull-down	0b000: OFF	0b11: INOUT
AL31	SCU_DSC	dsc_boot_mode, 2	VDD_ANA1_1P8 (1.8V)	Input	No init	0b10: Pull-down	0b000: OFF	0b11: INOUT
AJ29	SCU_DSC	dsc_boot_mode, 3	VDD_ANA1_1P8 (1.8V)	Input	No init	0b10: Pull-down	0b000: OFF	0b11: INOUT
AH30	M40_UART0	uart_tx	VDD_ANA1_1P8 (1.8V)	Output	0b000: Drive select 1mA	0b11: No pull	0b000: OFF	0b11: INOUT
AF28	M40_UART0	uart_rx	VDD_ANA1_1P8 (1.8V)	Input	No init	0b11: No pull	0b000: OFF	0b11: INOUT
AD28	SCU_JTAG	jtag_trst_b	VDD_ANA1_1P8 (1.8V)	Input	No init	0b11: No pull	0b000: OFF	0b11: INOUT
AF32	SCU_JTAG	jtag_tdo	VDD_ANA1_1P8 (1.8V)	Output	n/a	n/a	n/a	n/a
AG35	SCU_JTAG	jtag_tms	VDD_ANA1_1P8 (1.8V)	Input	n/a	n/a	n/a	n/a
AH34	SCU_JTAG	jtag_tdi	VDD_ANA1_1P8 (1.8V)	Input	n/a	n/a	n/a	n/a
AE31	SCU_JTAG	jtag_tck	VDD_ANA1_1P8 (1.8V)	Input	n/a	n/a	n/a	n/a
AG29	SCU_DSC	dsc_pmic_standby	VDD_ANA1_1P8 (1.8V)	Output	No init	0b11: No pull	0b000: OFF	0b11: INOUT
AR31	SNVS	snvs_pmic_on_req	VDD_SNVS_LDO_1P8_CAP (1.8V)	Output	n/a	n/a	n/a	n/a
AJ33	SCU_DSC	dsc_pmic_int_b	VDD_ANA1_1P8 (1.8V)	Input	0b000: Drive select 1mA	0b11: No pull	0b000: OFF	0b11: INOUT
AJ35	SCU_PMIC_I2C	pmic_i2c_scl	VDD_ANA1_1P8 (1.8V)	Input/Output	0b000: Drive select 1mA	0b11: No pull	0b000: OFF	0b10: OPEN_DRAIN_INPUT
AH32	SCU_PMIC_I2C	pmic_i2c_sda	VDD_ANA1_1P8 (1.8V)	Input/Output	0b000: Drive select 1mA	0b11: No pull	0b000: OFF	0b10: OPEN_DRAIN_INPUT

Clock Requirement for SoC

Normal run mode (S32K11x series)				
f_{SYS}	System and core clock	—	48	MHz
f_{BUS}	Bus clock	—	48	MHz
f_{FLASH}	Flash clock	—	24	MHz

Symbol	Parameter[1]	Value			Unit
		Min.	Typ.	Max.	
F_{FIRC}	FIRC target frequency	—	48	—	MHz
ΔF	Frequency deviation across process, voltage, and temperature < 105°C	—	±0.5	±1	$\%F_{FIRC}$
$\Delta F125$	Frequency deviation across process, voltage, and temperature < 125°C	—	±0.5	±1.1	$\%F_{FIRC}$
$T_{Startup}$	Startup time		3.4	5	μs^2
T_{JIT}[3]	Cycle-to-Cycle jitter	—	300	500	ps
T_{JIT}[3]	Long term jitter over 1000 cycles	—	0.04	0.1	$\%F_{FIRC}$

Clock requirements shall be defined for each block, so that we will not miss any configuration details in Software

Use-case	Description
VLPS and RTC	• Clock source: LPO or RTC_CLKIN
VLPS and LPUART TX/RX	• Clock source: SIRC • Transmiting or receiving continuously using DMA • Baudrate: 19.2 kbps
VLPS and LPUART wake-up	• Clock source: SIRC • Wake-up address feature enabled • Baudrate: 19.2 kbps
VLPS and LPI2C master	• Clock Source: SIRC • Transmit/receive using DMA • Baudrate: 100 kHz
VLPS and LPI2C slave wake-up	• Clock source: SIRC • Wake-up address feature enabled • Baudrate: 100 kHz
VLPS and LPSPI master [4]	• Clock source: SIRC • Transmit/receive using DMA • Baudrate: 500 kHz

Power Requirements for SoC

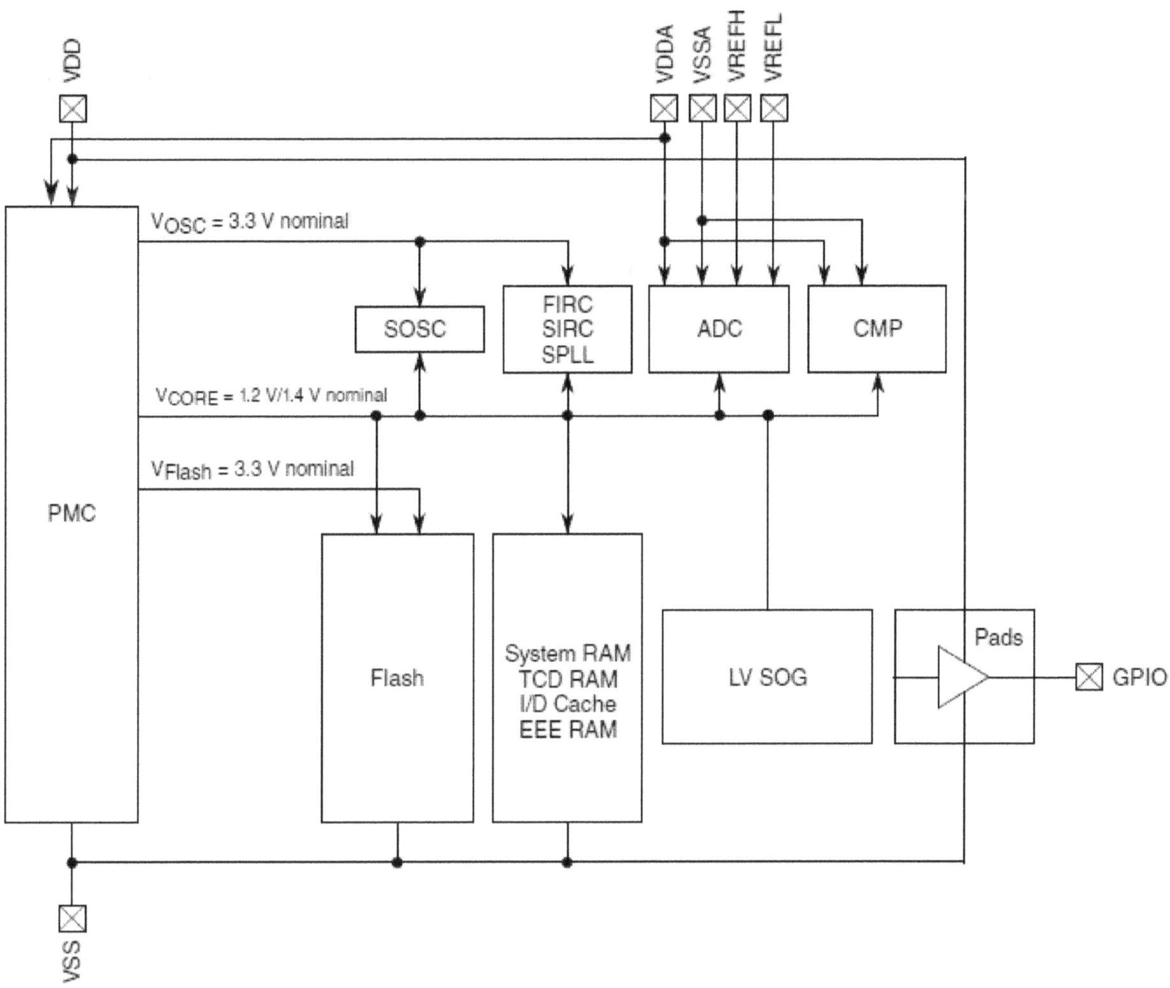

Figure 7. Power diagram

Refer 4th column in GPIO configuration table, power for each block and pin to be defined for current consumption analysis and power block failure analysis.

Peripheral Requirements for SoC

All the peripherals required for the application to be clearly defined.

Bit rate	Bit type	NBT(tq)	tq(ns)	SJW(tq)	TSEG2(tq)	Sampling point (%) Information only
500kbps	Normal	40	50	8	8	80%
2Mbps	Data	10	50	2	2	80%
500kbps	Normal	80	25	16	16	80%
2Mbps	Data	20	25	4	4	80%
500kbps	Normal	80	25	16	16	80%
5Mbps	Data	8	25	2	2	80%
500kbps	Normal	160	12.5	32	32	80%
5Mbps	Data	16	12.5	4	4	80%

HSI on Switches

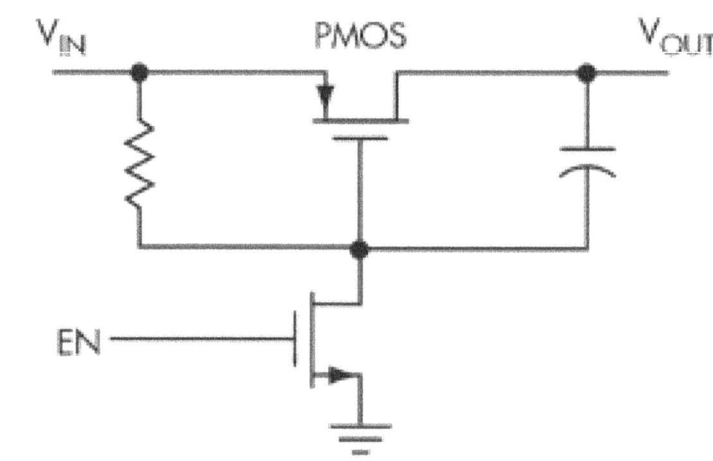

Load switches connect loads to the power supply under the control of an embedded microcontroller unit or Power Management control unit

They are many times used to control the sequence of power off-on operations. Since, the power sequence of a device can be critical for the microcontroller operation, interfaces wake-up, etc. It can also determine the activeness of a safety critical part in the operation.

The major parameters that are to be defined for a switch for safety analysis are:

- Control unit's Pin configuration

- Control unit's default configuration

- Input supply

- Output supply

Case Study for BTS5090 power supply switch

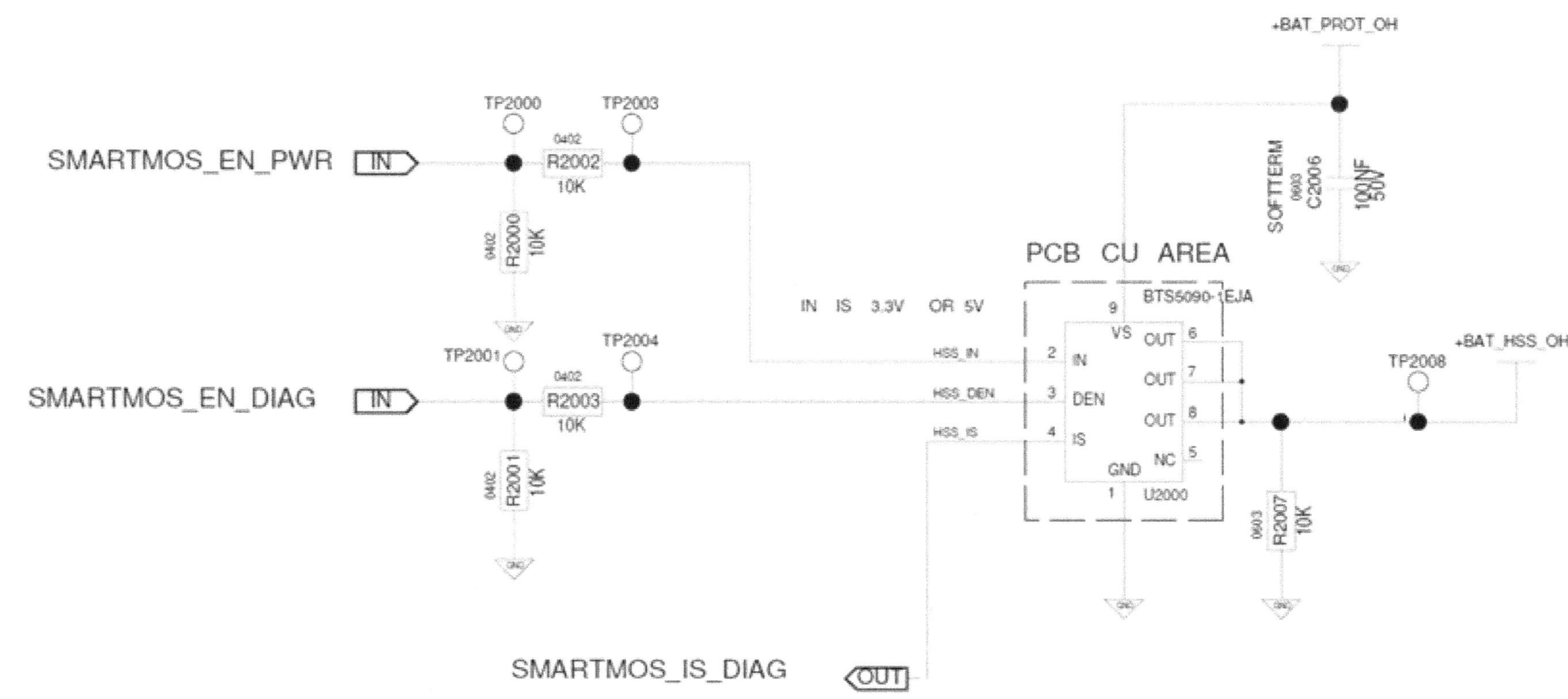

HSI on switches to be defined with voltage levels, necessary pins and circuitry

LABEL	PAD name	PAD N°	Power Group	Direction	Pull-Select	Output/Input
SMARTMOS_IS_DIAG	ADC_IN5	V34	VDD_ADC_DIG_1P8 (1.8V)	Input	0b11: No pull	0b11: INOUT
SMARTMOS_EN_PWR	USB_SS3_TC2	G15	VDD_USB_3P3 (3.3V)	Output	0b11: No pull	0b11: INOUT
SMARTMOS_EN_DIAG	USB_SS3_TC3	C15	VDD_USB_3P3 (3.3V)	Output	0b11: No pull	0b11: INOUT

HSI on Memories

Memories come under the safety-critical and essential element of a computing system because without it the ECU will fail to perform even simple tasks. RAM memory holds the programs and data that the MCU/MPU requires during the execution of a program. Read-only memory type stores crucial information essential to operate the system, like the program essential to boot the computer.

The Memories that have to be included in the safety-critical HSI documentation are:

- The internal Microcontroller memory

- DMA access configuration, if any

- External NAND/NOR FLASH memory, if used for safety-critical function

- External DDR memories, if used in safety-critical function

- Also, the small memories like EEPROM is included incase it's malfunctioning can affect any Safety Goal

The Memories section of HSI should contain register configuration details of each memory used in the system.

Example HSI provided for NOR Flash

BA2	BA1	BA0	A14-A13	A12	A11	A10	A9	A8	A7	A6	A5	A4	A3	A2	A1	A0	Address Field
0	0	0	0*1	PPD		WR		DLL	TM		CAS Latency		RBT	CL		BL	Mode Register 0

A8	DLL Reset
0	No
1	Yes

A7	mode
0	Nomal
1	Test

A3	Read Burst Type
0	Nibble Sequential
1	Interleave

A1	A0	BL
0	0	8 (Fixed)
0	1	BC4 or 8 (on the fly)
1	0	BC4 (Fixed)
1	1	Reserved

A12	DLL Control for Precharge PD
0	Slow exit (DLL off)
1	Fast exit (DLL on)

Write recovery for autoprecharge

A11	A10	A9	WR(cycles)
0	0	0	16*2
0	0	1	5*2
0	1	0	6*2
0	1	1	7*2
1	0	0	8*2
1	0	1	10*2
1	1	0	12*2
1	1	1	14*2

Minimum value

A6	A5	A4	A2	CAS Latency
0	0	0	0	Reserved
0	0	1	0	5
0	1	0	0	6
0	1	1	0	7
1	0	0	0	8
1	0	1	0	9
1	1	0	0	10
1	1	1	0	11
0	0	0	1	12
0	0	1	1	13
0	1	0	1	14
0	1	1	1	Reserved
1	0	0	1	Reserved
1	0	1	1	Reserved
1	1	0	1	Reserved
1	1	1	1	Reserved

BA1	BA0	MR Select
0	**0**	**MR0**
0	1	MR1
1	0	MR2
1	1	MR3

1. A14 and A13 must be programmed to 0 during MRS.
2. WR (write recovery for autoprecharge)min in clock cycles is calculated by dividing tWR(in ns) by tCK(in ns) and rounding up to the next integer: WRmin[cycles] = Roundup(tWR[ns] / tCK[ns]). The WR value in the mode register must be programmed to be equal or larger than WRmin. The programmed WR value is used with tRP to determine tDAL.
3. The table only shows the encodings for a given Cas Latency. For actual supported Cas Latency, please refer to speedbin tables for each frequency
4. The table only shows the encodings for Write Recovery. For actual Write recovery timing, please refer to AC timing table.

Figure 2.3.2 — MR0 Definition

HSI on Sensor

Sensors play an important role in automotives. The Hardware-Software Interface on sensors is critical for bridging the gap between the physical world and digital processing capabilities. It allows developers to harness the potential of sensor data and build applications ranging from simple data logging to complex real-time monitoring systems, automation, robotics, and more. Sensors monitor vehicle engines, fuel consumption and emissions, protecting drivers and passengers. These helps car manufacturers to launch cars that are much safer, fuel efficient and comfortable to drive.

Here's how the Hardware-Software Interface works on sensors:

Sensor Hardware: The physical sensor hardware is responsible for capturing data from the environment it is designed to monitor. Different types of sensors, such as image sensors, temperature sensors, pressure sensors, motion sensors, etc., have specific hardware components tailored to their respective sensing tasks. For instance, an image sensor might consist of an array of pixels to capture light intensity, while a temperature sensor could have a thermocouple or a thermistor to measure temperature changes.

Signal Processing and Conversion: The raw data obtained from the sensor hardware needs to be processed and converted into a digital format that the software can understand and work with. This may involve analog-to-digital conversion and other signal processing techniques to clean up noise and calibrate the sensor's readings.

Sensor Interface Protocol: To communicate with the software, sensors often use specific communication protocols or interfaces. Common examples include I2C (Inter-Integrated Circuit), SPI (Serial Peripheral Interface), UART (Universal Asynchronous Receiver/ Transmitter), or even standard communication protocols like USB (Universal Serial Bus) and Ethernet for more advanced sensors.

Device Drivers: In the software layer, device drivers act as intermediaries between the operating system and the sensor hardware. These drivers provide an interface for the software to interact with the sensor, allowing the software to send commands, receive data, and configure the sensor settings.

Sensor APIs and Libraries: Many sensors come with software development kits (SDKs), APIs, or libraries that offer higher-level abstractions for developers. These resources provide simplified functions and methods to interact with the sensor, reducing the complexity of low-level hardware interaction.

Sensor Data Processing: Once the sensor data is received by the software, it can be processed, analyzed, and used for various applications. This may involve performing computations, applying algorithms, or visualizing the data to extract meaningful information.

The Hardware-Software Interface on sensors is critical for bridging the gap between the physical world and digital processing capabilities. It allows developers to harness the potential of sensor data and build applications ranging from simple data logging to complex real-time monitoring systems, automation, robotics, and more.

HSI information on Sensor should contain two critical data:

- Sensor Resolution

- Numerical data expected for each real time value

ADC information		
Converter resolution	12	bits
Definition	ADC value	Voltage
ADC min:	0	0
ADC max:	4096	5

Case study of an NTC temperature sensor

Below is an example HSI of an NTC temperature sensor. In HSI, we calculate and determine the ADC value that will be read by the Microcontroller and what the value actually means. In addition to that, we define the range with tolerance of the NTC along with tolerance of the ADC of the Microcontroller.

Temperature (°C)	NTC_MIN (Ω)	NTC_TYP (Ω)	NTC_MAX (Ω)	VOUT_MIN	VOUT_TYP	VOUT_MAX	ADC_MIN	ADC_TYP	ADC_MAX	Temp_max[°C]	Temp_min[°C]
-40	183,680	191,910	200,130	4.5895969	4.7084565	4.8258130	3679	3858	4041	-26.64179104	-53.65671642
-35	140,770	146,590	152,420	4.5045819	4.6274199	4.7483102	3610	3791	3976	-23.6875	-46.5625
-30	108,730	112,880	117,030	4.4021366	4.5288438	4.6535212	3528	3711	3897	-20.56701031	-39.58762887
-25	84,620	87,588	90,556	4.2802601	4.4108445	4.5394062	3430	3614	3802	-16.92982456	-33.24561404
-20	66,339	68,471	70,602	4.1379034	4.2721261	4.4044460	3316	3500	3689	-12.97709924	-27.21374046

HSI on Power Management circuits

A power management integrated circuit (PMIC) is used to manage power on electronic devices or in modules on devices that may have a range of voltages. A Power Management Integrated Circuit (PMIC) is a specialized semiconductor device that is designed to manage and regulate the power supply for electronic systems and devices. It integrates various power-related functions into a single chip, optimizing power efficiency and simplifying the design of power delivery systems in electronic products.

PMICs are commonly used in a wide range of electronic devices, including smartphones, tablets, laptops, digital cameras, IoT devices, wearables, and other portable electronics. Their primary purpose is to efficiently regulate and distribute power to different components within the device to ensure stable and reliable operation while maximizing energy efficiency.

Power Management Integrated Circuits (PMICs) play a crucial role in managing and regulating the power supply for various electrical and electronic components within vehicles. As modern vehicles become more technologically advanced and incorporate numerous electronic systems, PMICs have become essential to ensure efficient and reliable power distribution and control. Here are some key areas where PMICs are used in automotive manufacturing:

Vehicle Control Units: PMICs are used in various control units within the vehicle, such as Engine Control Units (ECUs), Transmission Control Units (TCUs), and Body Control Modules (BCMs). These control units require stable and regulated power supplies to function correctly and communicate with other vehicle systems effectively.

Infotainment Systems: Modern automotive infotainment systems consist of multiple components, including displays, audio systems, navigation units, and connectivity modules. PMICs are used to regulate the power supply for these components, ensuring optimal performance and energy efficiency.

Advanced Driver Assistance Systems (ADAS): ADAS features, such as adaptive cruise control, lane departure warning, and collision avoidance systems, rely on various sensors and cameras. PMICs are used to power and manage the sensors and interface with the ADAS control units.

Lighting Systems: Automotive lighting systems, including headlights, taillights, and interior lighting, require precise power management. PMICs are utilized to control the brightness levels and ensure uniform illumination while optimizing power consumption.

Electric Vehicle (EV) Power Electronics: In electric and hybrid vehicles, PMICs are extensively used to manage the power flow between the battery, electric motor, and other electrical components. They help regulate the charging and discharging of the battery, as well as control the electric drivetrain efficiently.

On-Board Charging Systems: For electric vehicles, PMICs are employed in the charging systems to handle power conversion and charging control, ensuring safe and efficient charging of the vehicle's battery.

Power Distribution Modules: PMICs are used in power distribution modules that control and protect the flow of electrical power to different systems and circuits throughout the vehicle.

The use of PMICs in automotive manufacturing not only ensures optimal performance and reliability of the vehicle's electronic systems but also contributes to improved energy efficiency, reduced power losses, and enhanced safety. Additionally, the integration of power management functions into a single chip helps simplify the vehicle's electrical system design, reduce component count, and minimize the overall vehicle weight and size. As automotive technology continues to advance, PMICs are expected to play an even more significant role in powering the vehicles of the future.

In concern to Safety, an erroneous voltage/current regulation can lead to spurious functioning of all smart ICs in the circuits, say SOC, MCU, MPU, Memories, etc. Hence, it is critical to determine the controls and monitoring for power supply regulators which are supplying safety-critical components in the circuit.

Power supply IC used and the feedback path should be clearly defined, including the ADC or circuit used to monitor the power supply.

Case study of PMIC IC

LABEL	PAD name	PAD N°	Power Group	Direction	Pull-Select	Output/Input
PMIC_AMUX	ADC_IN2	V32	VDD_ADC_DIG_1P8 (1.8V)	Input	0b11: No pull	0b11: INOUT
PMIC_POR_B_1V8	POR_B	AG31	VDD_ANA1_1P8 (1.8V)	Input	n/a	n/a
SOC_PMIC_INT_B	PMIC_INT_B	AJ33	VDD_ANA1_1P8 (1.8V)	Input	0b11: No pull	0b11: INOUT
SOC_PMIC_ON_REQ	PMIC_ON_REQ	AR31	VDD_SNVS_LDO_1P8_CAP	Output	n/a	n/a
SOC_PMIC_SCL	PMIC_I2C_SCL	AJ35	VDD_ANA1_1P8 (1.8V)	Input/Output	0b11: No pull	0b10: OPEN_DRAIN_INPUT
SOC_PMIC_SDA	PMIC_I2C_SDA	AH32	VDD_ANA1_1P8 (1.8V)	Input/Output	0b11: No pull	0b10: OPEN_DRAIN_INPUT
SOC_PMIC_STBY	SCU_PMIC_STANDBY	AG29	VDD_ANA1_1P8 (1.8V)	Output	0b11: No pull	0b11: INOUT
SOC_JTAG_TRST_B	SCU_WDOG_OUT	AD28	VDD_ANA1_1P8 (1.8V)	Input	0b11: No pull	0b11: INOUT
PMIC_EWARN	CSI_PCLK	AK26	VDD_CSI_1P8_3P3 (1.8V)	Input	0b11: No pull	0b00: DEFAULT

Case study of the Monitoring Circuit

AMUX selection	Value (V)			Divider		Threshold		Unit	Value	
	Min	Typ	Max	Min	Max	Min	Max		Min	Max
VIN	2.50	5.00	5.50	3.97	4.05	0.617	1.385	V	1404	3152
VSNVS	2.85	3.00	3.15	3.48	3.54	0.805	0.905	V	1832	2059
LICELL	-	0.00	4.20	2.98	3.04	-	1.409	V	-	3207
SW1_FB	0.98	1.00	1.02	1.00	1.00	0.980	1.020	V	2230	2321
SW2_FB	0.98	1.00	1.02	1.00	1.00	0.980	1.020	V	2230	2321
SW3_FB	0.98	From 1,0 to 1,1	1.12	1.00	1.00	0.980	1.122	V	2230	2553
SW4_FB	1.08	1.10	1.12	1.00	1.00	1.078	1.122	V	2453	2553
SW5_FB	1.32	1.35	1.38	1.00	1.00	1.323	1.377	V	3010	3133
SW6_FB	0.66	0.68	0.69	1.00	1.00	0.662	0.689	V	1505	1566
SW7_FB	1.76	1.80	1.84	2.85	2.91	0.606	0.644	V	1379	1465

Chapter 8

Part 12 – Motorcycles

Topics to be covered

1. Part 12 of ISO26262

2. Vocabulary specific to Motorcycles

3. Adaptation of ISO26262 requirements for Two-wheelers

4. Adaptation of Safety culture

5. Re-classification of Confirmation measures

6. Modifications on HARA analysis

7. Motorcycle Safety Integrity Level (MSIL)

8. Case Study - Determination of MSIL

9. Modified methods for Vehicle Integration Testing

10. Modified methods for Safety Validation

11. Workproducts affected by Part 12 for Two-wheelers

12. Case Study - Goldwing runaway

13. Case Study - Malfunctions that lead to Motorcycle crash

Part 12 of ISO26262

The Concept of functional safety for automobiles was standardized in 2012 edition of ISO 26262 standard and it was only meant for road vehicles weighing up to 3500 kg. Even though the Motorcycles were not explicitly excluded in the first edition of the standard, the standard was essentially applicable only on cars or passenger vehicles.

177

The ISO26262 2018 version the 2nd edition of ISO26262 defines Part 12 especially for Two wheelers. The need to differentiate between the Safety analysis of other road vehicles and two wheelers is that this analysis places more responsibility on the motorcyclist rather than the motorcycle to mitigate risks.

Part 12 of ISO 26262 covers the "Adaptation of ISO 26262 for motorcycles." This part provides guidelines and adaptations of the standard's requirements specifically tailored to the safety aspects of motorcycles.

The inclusion of Part 12 acknowledges that motorcycles have unique characteristics compared to other road vehicles like cars and trucks. They have different dynamics, handling characteristics, and operational requirements. Therefore, the safety considerations and requirements need to be adapted accordingly to ensure the safe development and integration of electronics and electrical systems in motorcycles.

Please note that standards might undergo updates or revisions over time, and new parts or revisions may be published after my last update. Therefore, it is crucial to refer to the latest version of the ISO 26262 standard for the most current and accurate information, including any potential changes to Part 12 or other parts of the standard.

Why was Part 12 introduced in the 2018 version of ISO26262?

The reason why we consider that the Part 1-11 of ISO26262 does not cover motorcycles are as below:

- Similar components in a car and a bike would pose different hazards thus a need for a separate set of functional safety guidelines for motorcycles.

- The Three safety criticalities (Severity, Controllability and Exposure) work differently for Motorcycles.

- Due to the consideration of various parameters, the two-wheelers require different Functional safety guidelines.

- Few examples of such parameters are like the control available for the driver, the environments exposed, how the vehicle would react in case of an accident.

Difference Between Part 1-11 and Part 12 of ISO26262

Parameters	Part 1 - 11 of ISO26262	Part 12 of ISO26262
Scope of vehicles	Applicable from passenger cars up to heavy vehicles like buses and trucks (<3500kg)	Applicable only for Two-wheelers or three-wheelers (<800kg) except mopeds
Risk Classification	Automotive Safety Integrity Level (ASIL)	Motorcycle Safety Integrity Level (MSIL)
HARA	Safety goals and ASIL derived directly from HARA	Safety goals with MSIL derived from HARA and then MSIL is mapped to ASIL to follow ISO26262 requirements in parts 1-11
ASIL Rating to map ISO26262 development requirements	QM, A, B, C, D D is the highest ASIL	QM, A, B,C C is the highest ASIL (MSIL D)

Vocabulary specific to Motorcycles

Expert Rider – Role filled by persons capable of evaluating controllability classifications based on operation of actual motorcycles

Motorcycle – two-wheeled motor-driven vehicle, or three-wheeled motor-driven vehicle whose unladen weight does not exceed 800 kg, excluding mopeds as they follow ISO 3833 standard

Motorcycle Safety Integrity Level (MSIL) – One of four levels that specify the item's or element's necessary ISO 26262 risk reduction requirements. The MSIL is converted to ASIL for defining the safety measures to be satisfied for avoiding unreasonable residual risk for items and elements used for motorcycle applications, with MSIL D representing the most stringent and MSIL A the least stringent level

Controllability Classification Panel (CCP) – The assignment of the controllability class can be performed by a CCP, which can have expertise in the areas of:

- evaluation of motorcycle controllability [performed by expert rider(s)];

- motorcycle dynamics;

- electrical/electronic system;

- functional safety; or

- rider behavior.

Adaptation of ISO26262 requirements for Two-wheelers

Adapting ISO 26262 requirements for two-wheelers, specifically motorcycles, involves tailoring the functional safety standard to address the unique characteristics and safety considerations of these vehicles. Adaptation of ISO26262 requirements for motorcycles are the following:

- General topics for adaptation for motorcycles;

- Adaptation of safety culture;

- Reclassification of confirmation measures;

- hazard analysis and risk assessment;

- vehicle integration and testing; and

- safety validation.

General Topics for Adaptation for Motorcycles: This involves identifying and addressing the specific safety challenges and requirements that are specific to motorcycles. As motorcycles have different dynamics, handling characteristics, and operational scenarios compared to four-wheeled vehicles, certain aspects of ISO 26262 may need adjustments to suit the context of motorcycles.

Adaptation of Safety Culture: Establishing a safety culture is vital for any industry, including the motorcycle sector. This adaptation involves ensuring that manufacturers, suppliers, and all stakeholders in the motorcycle industry prioritize safety throughout the development and operational life cycle of the motorcycles.

Reclassification of Confirmation Measures: Confirmation measures refer to the processes and activities that validate and verify the safety functions of electronic and electrical systems in vehicles. Adapting these measures for motorcycles involves understanding the specific safety functions related to motorcycles and appropriately reclassifying them to align with the unique characteristics of two-wheeled vehicles.

Hazard Analysis and Risk Assessment: Hazard analysis and risk assessment are critical components of ISO 26262. For motorcycles, these processes need to consider the unique safety hazards associated with riding a two-wheeler, such as balance, maneuverability, and exposure to external factors like road conditions and weather.

Vehicle Integration and Testing: Adapting the vehicle integration and testing process for motorcycles involves ensuring that safety-related systems, including electronic and electrical components, are properly integrated into the motorcycle's design and thoroughly tested to meet safety requirements.

Safety Validation: Safety validation involves the process of verifying and validating that the safety requirements are met during the motorcycle's development and production. Adaptation in this context means ensuring the validation processes are tailored to address the specific safety aspects relevant to motorcycles.

Adapting ISO 26262 requirements for motorcycles is crucial to address the unique challenges and safety considerations associated with two-wheelers. It ensures that the safety of riders and other road users is adequately addressed throughout the development and operation of motorcycles, promoting a higher level of safety in the two-wheeler industry.

The classification requirement depends on the safety goals and their ASIL values. This is known as the "degree of independence" to be followed.

Degree of Independency

I0: The confirmation measure should be performed, if not justification needs to be provided. It shall be performed by a different person.

I1: The confirmation measure shall be performed by a different person.

I2: The confirmation measure shall be performed by a different person who is not reporting to the same direct superior.

I3: The confirmation measure shall be performed by a person from a different department or Organization, i.e, Independent from the department responsible for the considered work products regarding management, resources and release authority.

In Part-12 of the latest ISO 26262 standard, the confirmation measure has been re-classified for the motorcycle industry. I2 has been set as the highest level of independence requirement as compared to I3 in automotive functional safety. Also for motorcycles, Table 1 in Part 12 replaces the table in Part 2 defined for confirmation reviews for automotive functional safety. This table basically skips the ASIL D column as in the earlier table. ASIL D is missing in the table as the highest MSIL maps to ASIL C. Below is the table specific for motorcycles

Confirmation review on the Workproduct	Level of independence requirement			
	QM	ASIL A	ASIL B	ASIL C
Safety Impact Analysis	I3	I3	I3	I3
Hazard Analysis and Risk Assessment	I3	I3	I3	I3
Safety Development Plan	–	I1	I1	I2
Functional Safety Concept	–	I1	I1	I2
Technical Safety Concept	–	I1	I1	I2
Test Strategy for Integration testing	–	I0	I1	I2
Safety Validation Specification	–	I0	I1	I2
Safety Analysis (FMEDA, eFMEA, FTA)	–	I1	I1	I2
Dependent Failure Analysis	–	I1	I1	I2
Safety Case	–	I1	I1	I2
Functional Safety Audit	–	–	I0	I2
Functional Safety Assessment	–	–	I0	I2

Confirmation measures such as confirmation reviews and functional safety audits are to be performed satisfying the independence requirement defined for the highest ASIL defined for the system.

Modifications on HARA analysis

The main modification to HARA analysis is to define MSIL instead of ASIL. The few modifications made to the process of HARA are because of the reasons below:

- The safety in a two-wheeler ecosystem depends on multiple external factors such as helmets, protective gears, training, etc…

- The rider has an enhanced responsibility to keep the two-wheeler safe while riding.

- A hazard may also result from the motorcycle's behavior and not necessarily from a systematic failure.

Therefore the HARA analysis is more dependent on the rider and environment rather than the system alone.

Motorcycle Safety Integrity Level (MSIL)

The Motorcycle Safety Integrity Level (MSIL) is a metric used to assess the risk associated with safety hazards and define the necessary safety requirements for the motorcycle's electronic and electrical systems. MSIL is determined based on three factors: Exposure (E), Controllability (C), and Severity (S). Let's explore each factor:

1. **Exposure (E):** Exposure refers to the likelihood or probability of encountering a specific safety hazard during the motorcycle's operation. It assesses how often a hazardous situation may arise while using the motorcycle. For example, factors such as the frequency of specific road conditions, traffic patterns, and weather conditions can influence the exposure level.

2. **Controllability (C):** Controllability evaluates the degree to which a rider can control or mitigate the consequences of a potential hazard. It assesses the ability of the rider to respond to a hazardous situation effectively. For instance, factors such as motorcycle stability, maneuverability, and the effectiveness of safety features (e.g., brakes) contribute to the controllability factor.

3. **Severity (S):** Severity addresses the potential harm or severity of consequences resulting from a safety hazard. It evaluates the impact of the hazard on the rider, passengers, and other road users. Factors considered include the potential for injury, property damage, or loss of life in the event of an accident or system failure.

After assessing these three factors (E, C, and S), the MSIL is determined based on the combination of their values. The MSIL categorizes the motorcycle's safety integrity into discrete levels, with higher levels indicating a higher level of safety requirements. In general, the higher the MSIL level, the more stringent the safety measures and requirements applied to the motorcycle's electronic and electrical systems.

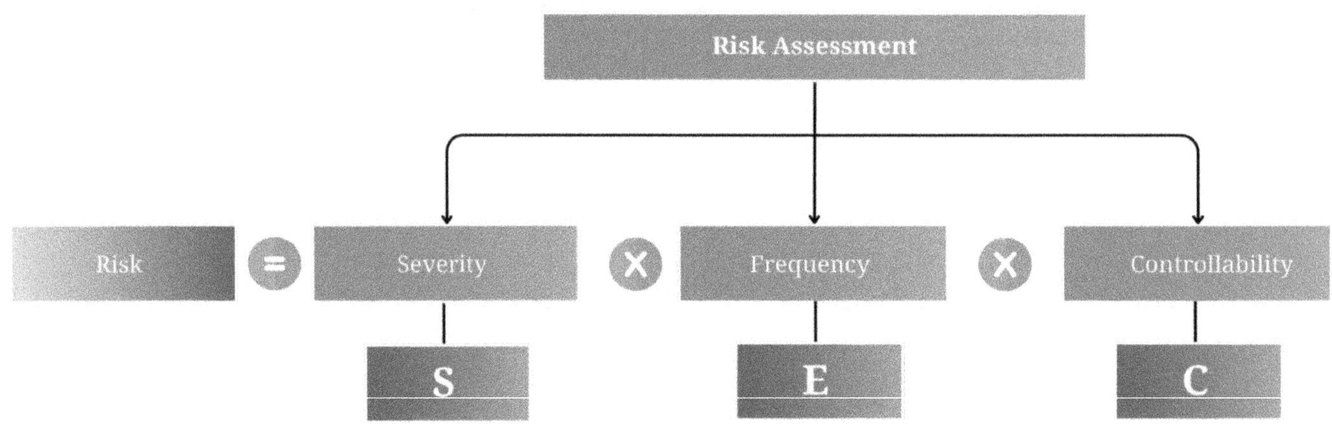

MSIL is derived from the factors Exposure (E), Controllability (C), and Severity (S), as mentioned earlier. Each factor is assessed on a scale of 1 to 4, and the MSIL value is determined based on the combination of these numerical ratings.

Here's a general representation of MSIL values based on the combination of factors:

1. MSIL 1: Low risk level - The combination of low exposure, high controllability, and low severity.

2. MSIL 2: Moderate risk level - The combination of moderate exposure, moderate controllability, and moderate severity.

3. MSIL 3: High risk level - The combination of high exposure, moderate to low controllability, and high severity.

4. MSIL 4: Very high risk level - The combination of very high exposure, low controllability, and very high severity.

It's essential to note that the specific criteria and numerical values used to assess Exposure, Controllability, and Severity may vary depending on the context and the safety standard or guidelines being followed. For example, ISO 26262, which is the standard for functional safety in automotive systems (including motorcycles), defines the criteria and evaluation methods for determining MSIL values.

When determining MSILs, there's also a fifth option — QM (quality management). This is used to indicate that the device that there isn't a safety requirement for that component.

However, it is generally still recommended that you comply with the requirement in order to improve the quality of the product.

- Identify the Item function of Item definition

- Identify Possible failure modes of the function

- Evaluate the functional failures in the relevant operational situations / operating modes

- Hazards caused by malfunctioning behaviour of the function shall be defined at the vehicle level.

- Classification of the hazards based on

    - Severity (S0,S1, S2, S3)

    - Exposure (E0,E1,E2,E3,E4)

    - Controllability (C0,C1,C2,C3)

- MSIL is determined based on the RISK matrix table

Table 5 — MSIL determination

Severity class	Exposure class	Controllability class		
		C1	C2	C3
S1	E1	QM	QM	QM
	E2	QM	QM	QM
	E3	QM	QM	A
	E4	QM	A	B
S2	E1	QM	QM	QM
	E2	QM	QM	A
	E3	QM	A	B
	E4	A	B	C
S3	E1	QM	QM	A
	E2	QM	A	B
	E3	A	B	C
	E4	B	C	D

MSIL to ASIL Mapping

- The output of HARA for a motorcycle is MSIL, the motorcycle counterpart of the ASIL.

- The method and the approach used to perform Hazard Analysis and Risk Assessment (HARA) for motorcycles is similar to that for the passenger vehicles.

- However, the ASIL-MSIL alignment/mapping has a important difference.

- For example, ASIL C for passenger cars is equivalent to MSIL D, the highest value for MSIL. Once the MSIL has been determined, it can be mapped to an equivalent ASIL

- A safety goal shall be determined for each hazardous event with an ASIL, mapped from MSIL, evaluated in the hazard analysis and risk assessment.

- Safety goals are most important safety requirements for the item. They lead to the functional safety requirements needed to avoid an unwanted risk for each hazardous event.

- Safety goals are not expressed in terms of technological solutions, but in terms of functional objectives.

Table 6 — Mapping of MSIL to ASIL

MSIL	ASIL
QM	QM
A	QM
B	A
C	B
D	C

- ISO 26262 Part-12 document provides Table-6 as the reference for mapping MSIL to ASIL. MSIL QM remains QM for ASIL, however, MSIL D is mapped to ASIL C.

- As per the standard, the ASIL levels mapped from the MSIL represent the minimum requirement. It implies that if the HARA determines the MSIL to be C, the component will be developed according to the requirements mentioned for ASIL B.

- However, to meet the requirements of any safety goal, the requirements mentioned in Part-12 will exceed the requirements in the other parts.

Case Study - Determination of MSIL

Example on how the ecosystem affects HARA

- A passenger car is designed to navigate safely through snow/ice on the road. Wheras, a motorcycle is not. So, if the rider decides to try off-roading or drives in a hazardous situation of heavy snowfall, he/she is accepting a higher degree of risk.

- Such a scenario is outside the scope of ISO 26262 Part-12. Moreover, the 3 important factors in Hazard Analysis and Risk Assessment

- HARA – controllability, severity and exposure are also affected to a great extent in such conditions.

Motorcycle specific hazard analysis and risk assessment

Motorcycle-specific hazard analysis and risk assessment are essential processes to ensure the safety of motorcycles and their riders. These assessments help identify potential hazards, analyze their consequences, and determine the level of risk associated with each hazard. It is more focused on rider behavior than the machine components, for mitigating risks. Controllability of motorcycle specific hazardous events place more focus on the rider.

The worldwide established level of technology in the motorcycle industry suggests that motorcycles safety cannot be handled with ASIL classification. So, an alignment between MSIL and ASIL classification is established to match ISO26262 to the worldwide capability of the motorcycle industry.

Identifying operating scenarios

- Malfunctions are considered in operational modes in the situations when the vehicle is correctly used and incorrectly used in a reasonably foreseeable way.

- Example: Road race, Motorcycle stunts, Motor cross or trial events are not considered normal motorcycle use conditions.

- HAZOP can be used to identify hazards and operational scenarios. Annex B lists severity scale based on AIS standard including the exposure (duration/frequency) probability examples.

- By taking into assumption that the driver is trained, experienced and in good condition.

- The scenarios should not be either too many or too few. The best way is to aggregate similar scenarios to the list 'as relevant as possible' to usage.

- Example: A normal motorcycle is not expected to travel on bad roads at high speed.

Derivation of ASIL for Charging a Battery

Assumption: The safety goals associated with a battery is a more critical consideration to be evaluated as per ASIL, more than the battery itself

The overcharging of battery at a speed below 10 km/hour is not as serious a situation as overcharging at very high speeds, the possibilities of overheating and consequent fire could also be high.

Vehicle Condition	Cause of malfunction	Possible hazard	ASIL
Running Speed< 10 km/h	Charging of battery pack beyond allowable energy storage	Overcharging may lead to thermal event	A
Running Speed> 10 – 50 km/h	Charging of battery pack beyond allowable energy storage	Overcharging may lead to thermal event	B
Running Speed> 50 km/h	Charging of battery pack beyond allowable energy storage	Overcharging may lead to thermal event	C

Modified methods for Vehicle Integration Testing

The changes in the ISO 26262 standard are not only confined to HARA and ASIL but also permeate to the testing activities.

Major modifications have been made in Product development at system level with respect to motorcycles.

There is a Table-7 for correct implementation of the functional safety requirements at the vehicle level. The test methods mentioned in the table will always get preference over the test methods defined in Part-4, Part-6 or Part-8 of the ISO 26262 standard (Only Motorcycle Functional Safety).

Table-8 gives the methods to ensure the correct functional performance, accuracy and timing of safety mechanisms at the vehicle level. These methods are recommended to fulfill the motorcycle-specific safety goals.

Table 7 — Correct implementation of the functional safety requirements at the vehicle level

Methods		ASIL		
		A	B	C
1a	Requirement-based test[a]	++	++	++
1b	Fault injection test[b]	++	++	++
1c	Long-term test[c]	++	++	++
1d	User test under real-life conditions[c,d]	++	++	++

[a] A requirements-based test denotes a test against functional and non-functional requirements.

[b] A fault injection test uses special means to introduce faults into the item. This can be done within the item via a special test interface or specially prepared elements or communication devices. The method is often used to improve the test coverage of the safety requirements, because during normal operation safety mechanisms are not invoked.

[c] A long-term test and a user test under real-life conditions are similar to tests derived from field experience but use a larger sample size, normal users as testers, and are not bound to prior specified test scenarios, but performed under real-life conditions during everyday life. These tests can have limitations if necessary to ensure the safety of the testers, e.g. with additional safety measures or disabled actuators. Long-term tests can be infeasible for motorcycles.

[d] User tests can be infeasible for motorcycles.

Table 8 — Correct functional performance, accuracy and timing of safety mechanisms at the vehicle level

Methods		ASIL		
		A	B	C
1a	Performance test[a]	+	+	++
1b	Long-term test[b]	+	+	++
1c	User test under real-life conditions[b,c]	+	+	++
1d	Fault injection test[d]	o	+	++
1e	Error guessing test[e]	o	+	++
1f	Test derived from field experience[f]	o	+	++

[a] A performance test can verify the performance (e.g. fault tolerant time intervals on vehicle level and vehicle controllability in the presence of faults) of the safety mechanisms concerning the item.

[b] A long-term test and a user test under real-life conditions are similar to tests derived from field experience but use a larger sample size, normal users as testers, and are not bound to prior specified test scenarios, but performed under real-life conditions during everyday life. These tests can have limitations if necessary to ensure the safety of the testers, e.g. with additional safety measures or disabled actuators. Long-term tests can be infeasible for motorcycles.

[c] User tests can be infeasible for motorcycles.

[d] A fault injection test uses special means to introduce faults into the item. This can be done within the item via a special test interface or specially prepared elements or communication devices. The method is often used to improve the test coverage of the safety requirements, because during normal operation safety mechanisms are not invoked.

[e] An error guessing test uses expert knowledge and data collected through lessons learned to anticipate errors in the system. Then a set of tests along with adequate test facilities is designed to check for these errors. Error guessing is an effective method given a tester who has previous experience with similar systems.

[f] A test derived from field experience and data gathered from the field.

Modified methods for Safety Validation

Safety validation covers:

- the controllability (including intended use and foreseeable misuse)

- the effectiveness of the external measures

- the effectiveness of the elements of other technologies (For example, a mechanical component that prevents a malfunction can be validated on the final vehicle at a later stage)

Correct implementation of internal and external interfaces at the vehicle level

	Methods	ASIL		
		A	B	C
1a	Test of internal interfaces[a]	+	+	++
1b	Test of external interfaces[a]	+	+	++
1c	Test of interaction/communication[b]	+	+	++
[a]	An interface test at the vehicle level tests the interfaces of the vehicle systems for compatibility. This can be done statically by validating value ranges, ratings or geometries as well as dynamically during operation of the whole vehicle.			
[b]	A communication and interaction test includes tests of the communication between the systems of the vehicle during runtime against functional and non-functional requirements.			

Level of robustness at the vehicle level

Methods		ASIL		
		A	B	C
1a	Resource usage test[a]	+	+	++
1b	Stress test[b]	+	+	++
1c	Test for interference resistance and robustness under certain environmental conditions[c]	+	+	++
1d	Long-term test[d]	+	+	++

[a]   At the vehicle level, resource usage testing is usually performed in dynamic environments (e.g. electronic control unit network environments, prototypes or whole vehicles). Issues to test include item internal resources, power consumption or limited resources of other vehicle systems.

[b]   A stress test verifies the correct operation of the vehicle under high operational loads or high demands from the environment. Therefore tests under high loads on the vehicle or with extreme user inputs or requests from other systems as well as tests with extreme temperatures, humidity or mechanical shocks can be applied.

[c]   A test for interference resistance and robustness, under certain environmental conditions, is a special case of stress testing. This includes EMC and ESD tests (e.g. see References [4] and [5]).

[d]   A long-term test and a user test under real-life conditions are similar to tests derived from field experience but use a larger sample size, normal users as testers, and are not bound to prior specified test scenarios, but performed under real-life conditions during everyday life. Long-term tests can be infeasible for motorcycles.

- The safety goals shall be validated for the item in a representative context at vehicle level

- Integrated item includes system, software, hardware, elements of other technologies, external measures as applicable.

- Appropriate vehicle types and vehicle configuration shall be considered

- HARA analysis report provides sources of information regarding relevant input for the choice of representative vehicles

The achievement of functional safety for the item when being integrated into the vehicle shall be validated by evaluating the following aspects

a. the controllability;

Controllability can be validated using operating scenarios, including intended use and foreseeable misuse. One acceptance criterion for the safety validation might be a sufficient controllability in a safe state defined in ISO 26262-3:2018, 7.4.2.5. A single acceptance criterion might not be sufficient to verify a safe state.

b. the effectiveness of the external measures;

c.   the effectiveness of the elements of other technologies;

d.   assumptions that influence the ASIL mapped from MSIL in the hazard analysis and risk assessment

The safety validation at the vehicle level, based on the safety goals, the functional safety requirements and the intended use, shall be executed as planned using, Operational use cases can be created to help focus the safety validation at the vehicle level.

An appropriate set of the following methods shall be applied:

a.   repeatable tests with specified test procedures, test cases, and pass/fail criteria;

EXAMPLE 1 Positive tests of functions and safety requirements, black box testing, simulation, tests under boundary conditions, fault injection, durability tests, stress tests, highly accelerated life testing (HALT), simulation of external influences.

b.   analyses; - EXAMPLE 2 FMEA, FTA, ETA, simulation.

c.   long-term tests, - such as vehicle driving schedules and captured test fleets; Long-term tests with targeted users can be infeasible for motorcycles.

d.   user tests under real-life conditions, panel or blind tests, expert panels;

User test can be infeasible for motorcycles. Real-life condition can be conducted using simulated condition

e.   reviews.

Work products affected by Part 12 for Two-wheelers

Work products by Clause 8 and Clause 10

- Hazard analysis and risk assessment report resulting from requirements

- Verification report of the hazard analysis and risk assessment resulting from requirements

- Safety validation specification including safety validation environment description

- resulting from requirements

- Safety validation report resulting from requirements

Case Study - Goldwing runaway

Honda 1800cc Goldwing Runaway[2]

Honda's luxurious Gold Wing is a two wheeler with flat six-cylinder engine and 7-speed Dual Clutch Transmission option. Goldwing has airbag safety option.

On 2004, an accident happens where the victim claims to have uncontrollable race of the engine for the malfunction. As soon as the engine starts, the bike moves at max throttle and moves for 58 feet, hits the curb before the victim is thrown off.

The victim has minor injuries as it was a very low traffic zone and gets a new bike as insurance proceeds. Affected party - The OEM is roasted legally and on media, faces loss of brand's name and loyal customers.

Case Study - Malfunctions that lead to Motorcycle crash

Not even the most experienced motorcycle riders can be entirely confident that they won't crash. Even when they have taken proper safety measures, their vehicles may malfunction and lead to fatalities

Motorcycle crashes can occur due to various malfunctions and failures that compromise the vehicle's safety and control. Here are some common malfunctions that can lead to motorcycle crashes:

- Brake Failure: Brake system malfunctions can result in reduced or complete loss of braking power, making it difficult for the rider to slow down or stop the motorcycle effectively.

- Tire Blowouts: Defective or worn-out tires can experience sudden blowouts, causing the rider to lose control of the motorcycle and potentially leading to a crash.

- Suspension Issues: Faulty suspension can affect the motorcycle's stability, leading to difficulty in handling and increased risk of accidents, especially in uneven or bumpy road conditions.

- Electrical Problems: Electrical malfunctions can impact critical systems such as lights, turn signals, or engine control, leading to reduced visibility and impaired performance.

- Steering Problems: Issues with the steering system, such as misalignment or malfunctioning, can affect the rider's ability to control the motorcycle's direction.

- Throttle Sticking: A throttle that sticks in the open position can cause sudden acceleration, leading to loss of control and potential crashes.

- Fuel System Failures: Problems in the fuel system, such as leaks or blockages, can cause engine stalls or fires, posing serious safety risks.

- Clutch Malfunction: A malfunctioning clutch can affect gear shifting and lead to difficulty in controlling the motorcycle's speed and acceleration.

- Chain or Belt Issues: The motorcycle's chain or belt transfers power from the engine to the rear wheel. A broken or loose chain/belt can lead to sudden loss of power to the rear wheel.

- Frame or Structural Flaws: Structural integrity issues in the motorcycle's frame can compromise its stability and safety.

- Ignition System Problems: Faulty ignition systems can cause engine misfires or prevent the motorcycle from starting altogether.

- Exhaust System Failures: Exhaust leaks or other issues can lead to toxic fumes entering the rider's cabin and impairing their ability to concentrate.

It's crucial for motorcycle owners and riders to perform regular maintenance checks and inspections to identify and address potential malfunctions promptly. Routine maintenance and servicing can go a long way in preventing accidents caused by mechanical failures. Additionally, rider training and awareness of motorcycle safety best practices are essential to minimize the risk of crashes due to handlebar errors or other rider-related factors.

Chapter 9

SOTIF – Safety of the Intended Functionality

Topics to be covered

1. Definition of SOTIF

2. Why is SOTIF needed?

3. Difference between SOTIF(ISO21448) and FuSa(IEC61508)

4. SOTIF principles

5. SOTIF Hazard Identification

6. SOTIF workflow

7. Case Study - SOTIF and ADAS

8. Case Study - Safe autonomous driving

Definition of SOTIF

The practical operation of automated vehicles in a real-world environment requires that they achieve and maintain a certifiable functionally safe state. These vehicles must maintain this state even in the chaotic environments shared with older vehicles that may have less or no automation.

"SOTIF," an acronym which stands for "Safety Of The Intended Functionality." By definition, it is the absence of unreasonable risk due to hazards resulting from functional insufficiencies of the intended functionality or from reasonably foreseeable misuse by persons. The aim is that the function can be controlled in all situations in relation to their operation and can be managed by the user in a safe manner. Thereby concluding the reasonable risk can be tolerated. The current state of the standard is defined under ISO/DIS 21448:2021

SOTIF deals with the safety of a system's behavior, particularly in situations where no component failure or malfunction occurs, but there are potential hazards due to the system's design and intended functionality. In other words, SOTIF addresses situations where the system behaves as intended but might lead to hazardous situations or risks due to unforeseen scenarios, human-machine interactions, or environmental factors.

Key aspects of SOTIF include:

Foreseeable Misuse: Evaluating the system's behaviour under reasonably foreseeable misuse scenarios and ensuring that it does not create hazardous situations.

- External Factors: Considering external factors like weather conditions, road infrastructure, and human behaviors that could interact with the system and potentially lead to safety risks.

- Operational Design Domains (ODD): Defining the operational design domains in which the system is intended to operate safely and identifying potential hazards outside those domains.

- Risk Assessment: Conducting risk assessments to identify potential hazards arising from the intended functionality and determining necessary measures to mitigate those risks.

- Validation and Verification: Ensuring that the system's intended functionality is accurately defined, and validating its safety under a wide range of scenarios through simulations, tests, and real-world evaluations.

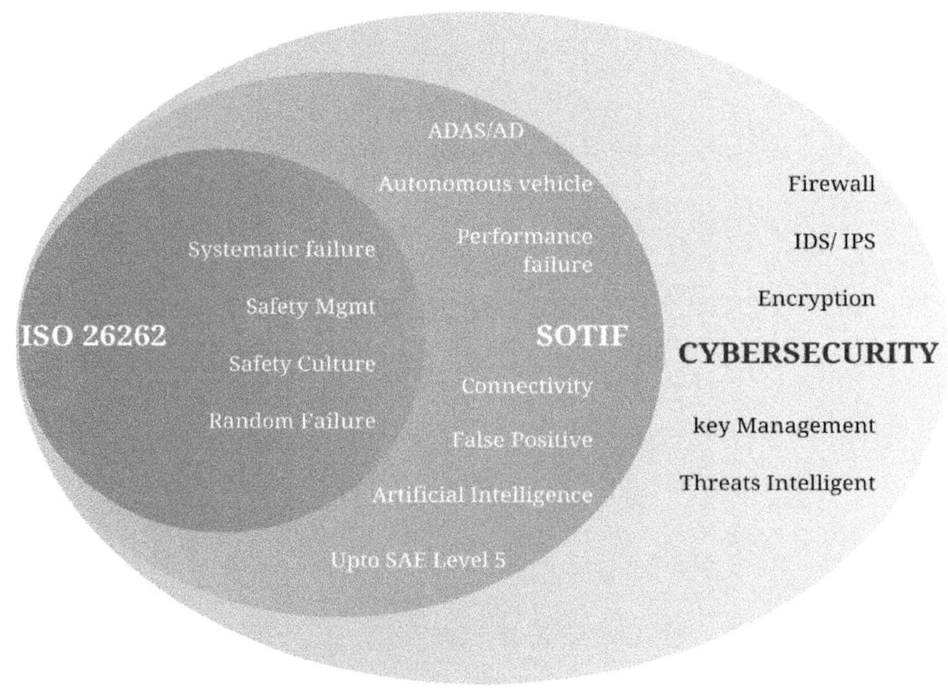

The situational awareness is derived from complex sensors and processing algorithms; especially emergency intervention systems such as interference of emergency braking systems and ADAS levels 1 and 2. It is also applicable to innovative functions if situational awareness from complex sensors and processing algorithms are part of the innovation.

SOTIF is complementary to functional safety (ISO 26262) and focuses on aspects beyond the system's hardware and software safety. It plays a crucial role in addressing potential hazards arising from the system's intended behavior and interactions with the environment, thus contributing to the overall safety of advanced automotive technologies, including ADAS and autonomous vehicles.

SOTIF aims to reduce threats to automotive safety such as:

- Residual risk of the intended function through analysis

- Unintended behavior in known situations through verification

- Residual unknown situations that could cause unintended behavior through validation of verification scenarios

Why is SOTIF needed?

SOTIF is needed to complement traditional functional safety practices and address the broader safety aspects of advanced vehicle technologies. By considering intended system behavior, human-machine interactions, and the dynamic driving environment, SOTIF enhances the overall safety of ADAS and autonomous vehicles, paving the way for safer and more reliable transportation systems. Here are the key reasons why SOTIF is necessary:

Addressing Unintended Hazards: SOTIF focuses on identifying and mitigating hazards that arise from the intended behavior of the system. It addresses scenarios where the system operates as designed but might encounter unexpected or rare situations that could lead to safety risks. Traditional functional safety standards like ISO 26262 mainly deal with failures and malfunctions, but SOTIF complements this by considering the safety implications of correct system functioning.

Complexity of ADAS and Autonomous Systems: ADAS and autonomous vehicle technologies are becoming increasingly complex, involving sophisticated algorithms, sensor fusion, and decision-making processes. As these systems interact with a dynamic environment and human drivers, the potential for unforeseen scenarios and risks increases. SOTIF ensures a more comprehensive approach to safety by considering the system's interactions and the environment it operates in.

Human-Machine Interaction: SOTIF recognizes the importance of human factors in the operation of advanced vehicle systems. It addresses potential hazards resulting from human-machine interactions, such as drivers misusing or overrelying on ADAS features, leading to unsafe behaviors. By considering human factors, SOTIF aims to design systems that are safe and intuitive for users.

Operational Design Domains (ODD): SOTIF emphasizes defining the operational design domains in which the system is intended to operate safely. By clarifying the system's limitations and intended usage, SOTIF helps prevent potential hazards that may arise when operating outside these boundaries.

Preventing Incidents: SOTIF aims to proactively prevent incidents and accidents that could occur due to unforeseen interactions between the system and the environment. By identifying and mitigating potential hazards during the development phase, it reduces the likelihood of incidents in real-world driving scenarios.

Regulatory and Industry Requirements: As the automotive industry progresses towards higher levels of automation, regulators and stakeholders demand a holistic approach to safety. SOTIF is increasingly becoming a part of safety standards and guidelines to ensure that vehicles are not only functionally safe but also operate safely in their intended use cases.

The Modern Mechatronic System: SOTIF (Safety Of The Intended Functionality) is essential for modern mechatronic systems due to the increasing complexity and integration of mechanical, electrical, and software components in these systems. Mechatronic systems are prevalent in various industries, including automotive, aerospace, robotics, and industrial automation.

Difference between SOTIF(ISO21448) and FuSa(ISO26262)

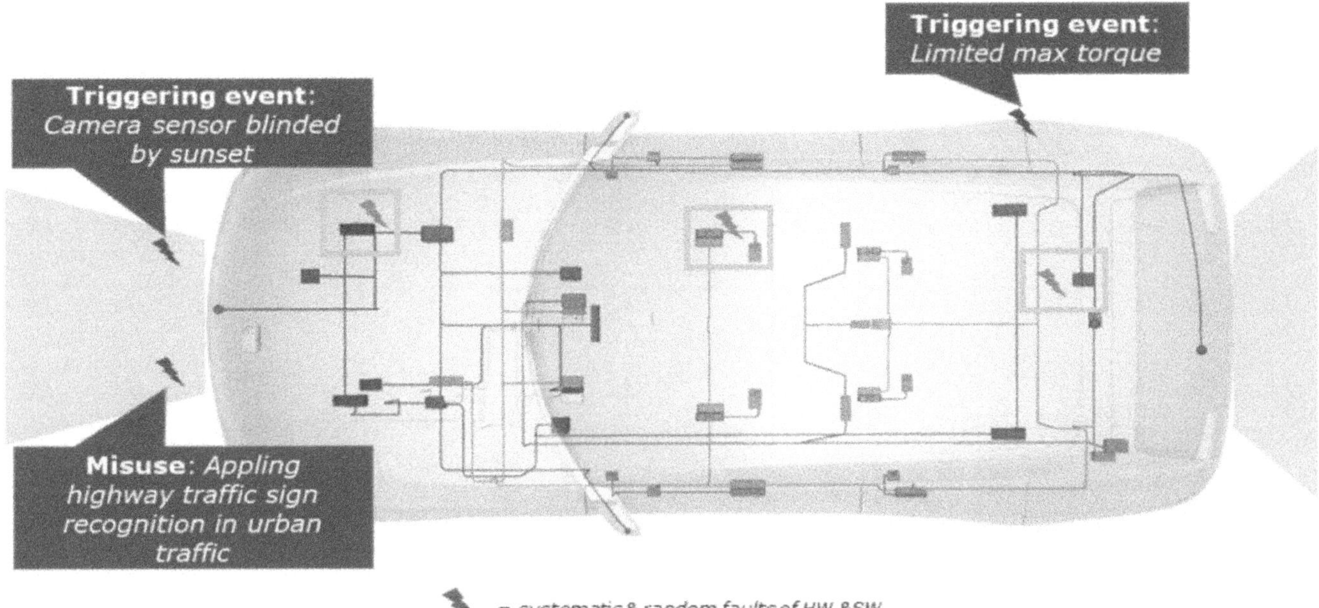

ISO26262 (FuSa)	ISO 21448 (SOTIF)
This is a risk-based safety standard which defines functional safety for automotive equipment, applicable throughout the life cycle of all electronic and electrical safety-related systems, ranging from the specification, to design, implementation, integration, verification, validation, and production release.	This standard is specific for Safety Of The Intended Functionality (SOTIF). SOTIF is applicable to analyze the insufficiencies of the intended functionality, or by reasonably foreseeable misuse by persons. The standard ISO 21448 provides guidance on the applicable design and verification and validation measures needed to achieve the SOTIF. It does not apply to cases covered by the ISO 26262 or to hazards directly caused by the system technology.

SOTIF principles

SOTIF-related hazardous event model

The main objective is to describe the process and rationale used to ensure that the risk level associated with a SOTIF-related hazardous event is sufficiently low. The function, system specification and design include relevant use cases which, in turn, are comprised of several relevant scenarios. These scenarios could contain triggering conditions that lead to harm. In order to avoid the harm, proper situational awareness may be necessary.

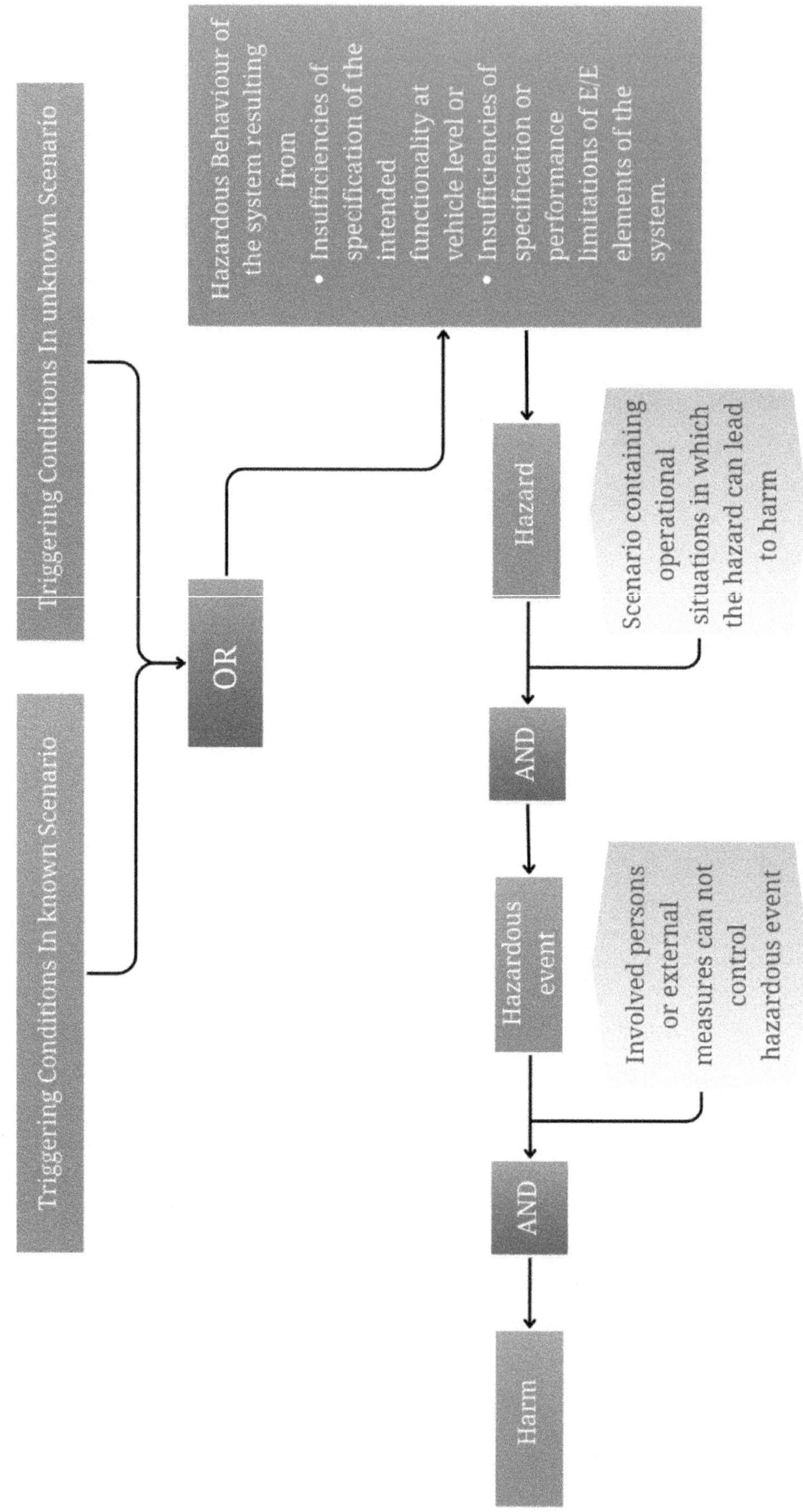

Triggering Conditions In unknown Scenario
Triggering Conditions In known Scenario
OR
Hazardous Behaviour of the system resulting from
Insufficiencies of specification of the intended functionality at vehicle level or
Insufficiencies of specification or performance limitations of E/E elements of the system.
Hazard
Scenario containing operational situations in which the hazard can lead to harm
AND
Hazardous event
Involved persons or external measures can not control hazardous event
AND
Harm

a Trigger conditions includes reasonably foreseeable direct misuse.

b The inability to control the hazardous event can also be the result of a reasonably foreseeable indirect misuse, e.g. the driver does not supervise the system as he is supposed to do.

Example 1 Activating a functionality intended for the highway in an urban setting has difficulties in recognizing and interpreting the motion of vulnerable road users.

Example 2 Confusion of the system operating mode such as the driver who assumes that the system is active even though it is deactivated. In such a situation, the potential insufficiencies of the system HMI to prevent from this confusion or the absence of an appropriate system reaction (if the driver behaviour can be monitored) can also be considered as a hazardous behaviour of the system.

For the proper situational awareness it is critical to have:

- Sufficiently comprehensive and accurate perception of the relevant environmental conditions and a correct understanding of the scene (e.g. detecting a relevant stop sign)

- Knowledge of appropriate actions or reactions in the driving scene (e.g. obey rules associated with stop signs)

Over the vehicle operational life, the following items can vary:

- The environment (e.g. new type of traffic signs, road markings, vehicles, …)

- Appropriate reactions (e.g. new driving action required by a new traffic sign; changes in driving scenarios, changes in driving laws, etc.)

Such considerations are taken into account when specifying the Operational Design Domain and during system development (risk identification, definition of appropriate measures) to ensure the SOTIF during operation

The four areas scenario categories

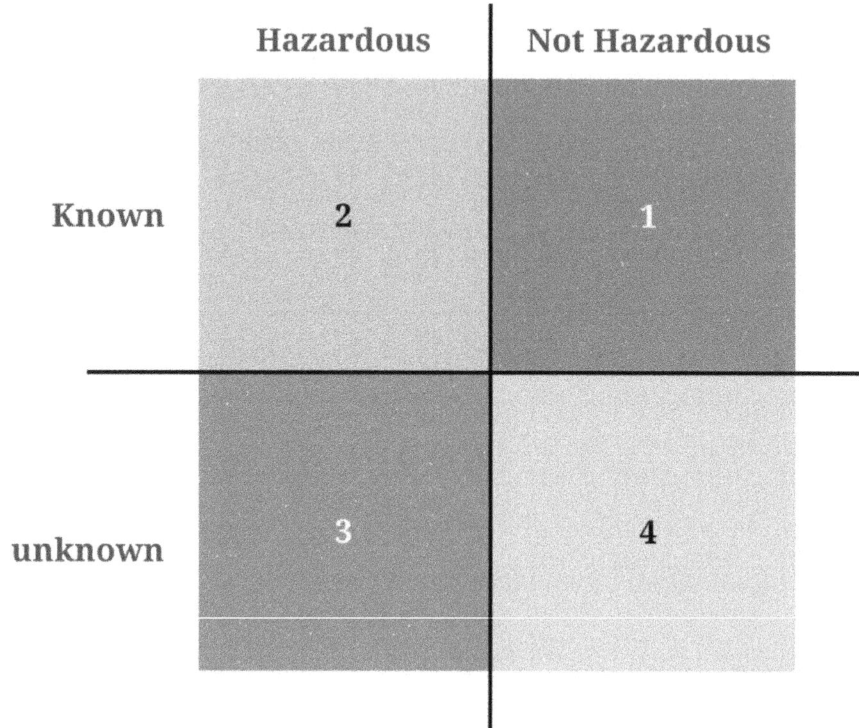

The scenario analysis in SOTIF is divided into four categories:

- Known not hazardous scenarios (Area 1);

- Known hazardous scenarios (Area 2);

- Unknown hazardous scenarios (Area 3); and

- Unknown not hazardous scenarios (Area 4).

This model is a conceptual abstraction representing a goal of the SOTIF process, which is to:

- Perform a risk acceptance evaluation of Area 2 based on the analysis of the intended functionality.

- Reduce the probability of known scenarios causing hazardous behaviour, in Area 2, to an acceptable level of risk.

- Reduce the probability of the unknown scenarios causing potentially hazardous behaviour, in Area 3, to an acceptable level of risk.

- Scenarios in Area 4 that are unknown but not hazardous do not impose risk. Once a scenario in Area 4 is discovered (i.e. becomes known), it is moved to Area 1.

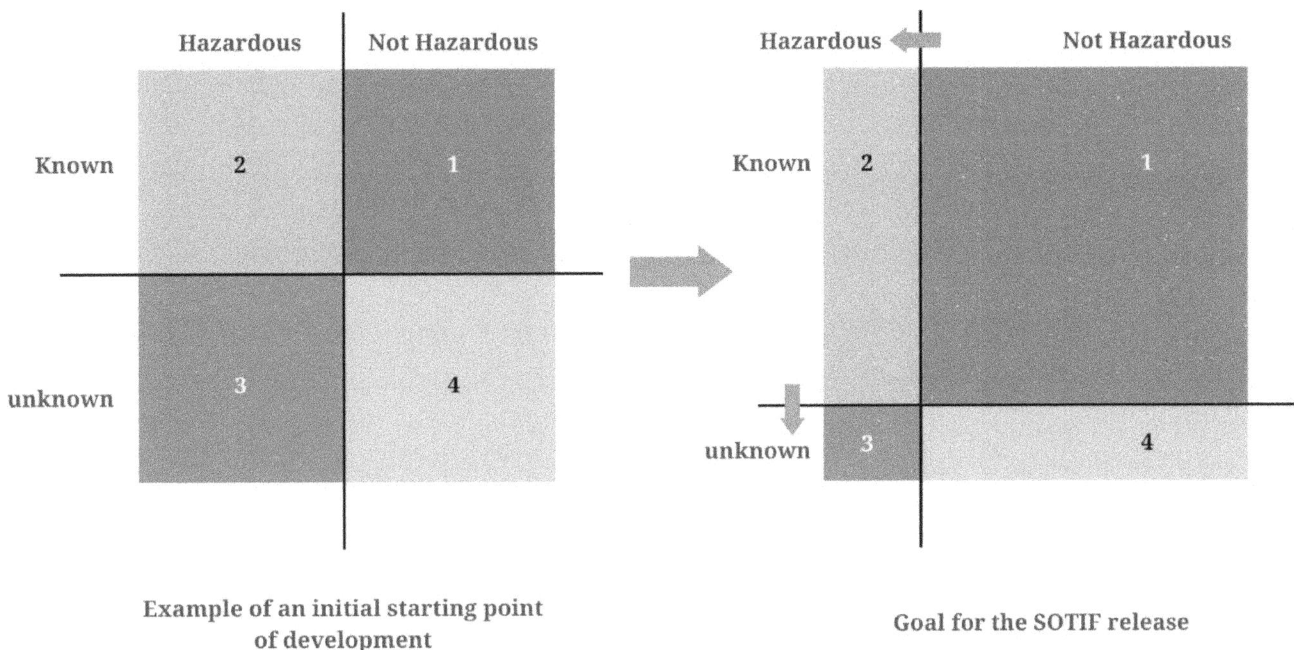

Example of an initial starting point of development

Goal for the SOTIF release

SOTIF Hazard identification

The SOTIF requirements for any functional system will be derived by the below defined steps:

Deriving SOTIF requirements

The Function and System Specification has two normative requirements

- Compile and create evidence containing the information sufficient to initiate the SOTIF related activities

- Update the evidence as necessary after each iteration of the SOTIF related activities

Defining from Functional Specifications

a. The goals of the intended functionality

b. The use cases in which the intended functionality is activated, deactivated and is active

c. The description of the intended functionality

d. The level of automation/authority over the vehicle dynamics

e. The dependencies on, and interaction with:

   a. the car driver, passengers, pedestrians and other road users

   b. relevant environmental conditions

   c. the interfaces with the road infrastructure

Defining from System Specifications

1. The description of the system and elements implementing the intended functionality

2. The description and behaviour of the installed sensors, controllers and actuators used by the intended functionality

3. The assumptions about how the intended functionality makes use of inputs from other elements

4. The assumptions about how other elements make use of outputs from the intended functionality

5. The concepts and technologies for the system and sub systems

6. The limitations and their countermeasures

7. The system architecture supporting the countermeasures

8. The degradation concept

9. The warning strategies

10. The dependencies on, and interaction with other functions and systems of the vehicle

SOTIF workflow

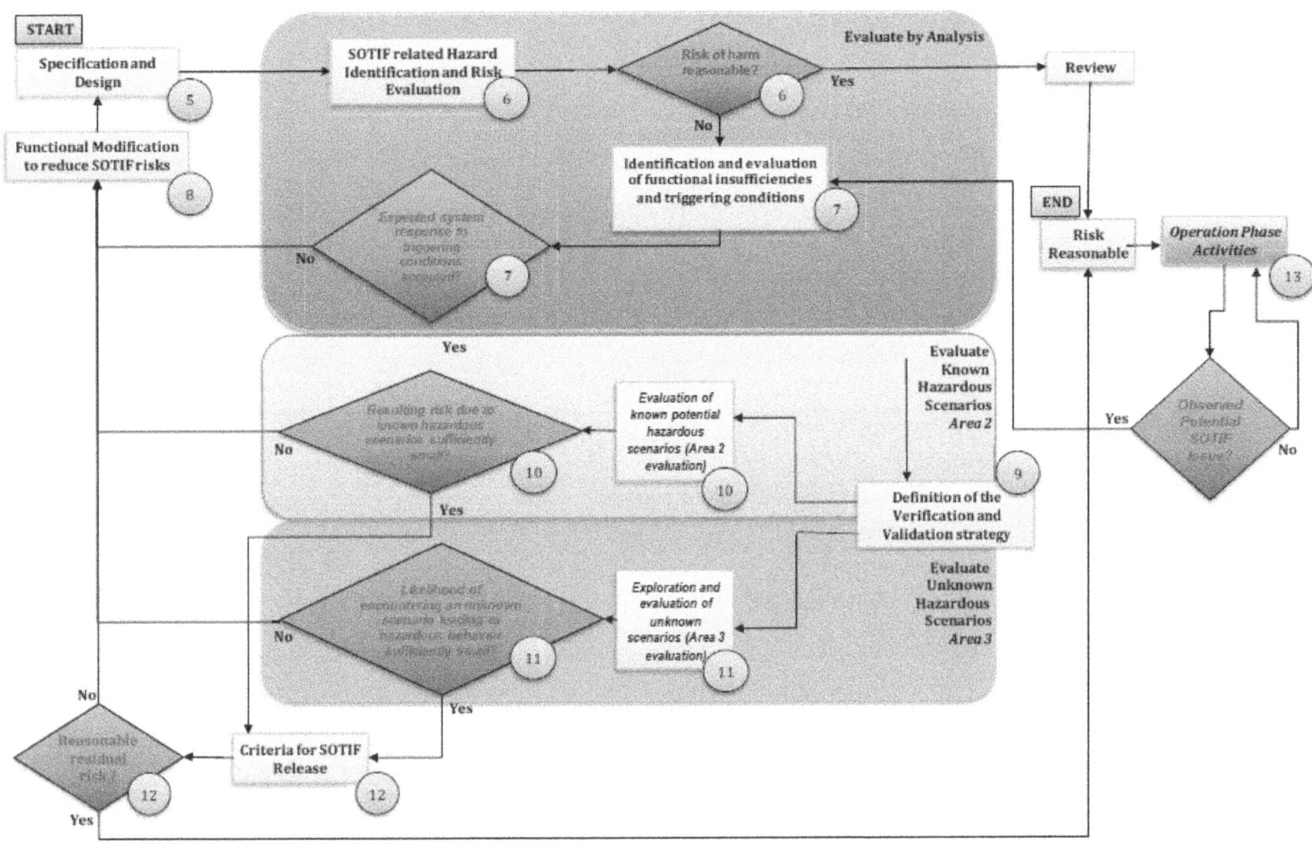

Note: The number are sections from ISO/PAS 21448:2019 'Road vehicles – Safety of the intended functionality' (SOTIF)

5 - Specification and Design

The specification and design can include various aspects relevant for a specific automation level or a specific implementation.

6 - Identification and evaluation of Hazards

The hazards resulting from functional insufficiencies are determined systematically at the vehicle level. This systematic identification is primarily based on knowledge about the function and its possible deviations resulting from functional insufficiencies. This can be achieved by applying the methods specified in ISO 26262-3:2018.

7 - Identification and Evaluation of Functional insufficiencies and Triggering conditions

A systematic method can be established to perform the analysis of functional insufficiencies and triggering conditions. This method can consider knowledge gained from similar projects, experts and field experience. The analysis aims to identify the insufficiencies of specification, the performance limitations (including those of sensors, decision algorithms and actuators) and the related scenarios (containing triggering conditions) that could lead to harm.

8 - Functional Modification and SOTIF risks

The conditions under which the activities in this clause consider measures addressing SOTIF related risks, are as follows:

- The intended functionality of the initial specification and design is identified as having a hazardous scenario that requires further analysis, i.e. assessed as likely to cause harm in the risk assessment of the hazardous event; and

- That the system's response to the identified triggering condition that causes the hazardous scenario that is deemed unacceptable i.e. there is a known scenario where the residual risk of triggering the hazardous event does not meet the acceptable criteria and leads to an unreasonable risk.

Under the above conditions, the system is elaborated through the iteration of considering SOTIF measures in this clause, updating the specification and design with these SOTIF measures, and risk assessment of the intended functionality is conducted by using the updated specification and design.

This elaborated system (including the effectiveness of the SOTIF measures) will then be evaluated in the Verification & Validation phase.

9 - Definition of Verification and Validation Strategy

The system verification and validation activities regarding the risk of potentially hazardous behaviour(excluding the faults addressed by ISO 26262) include integration testing activities to address the following scope:

1. The ability of sensors to provide accurate information on the environment to meet the performance requirements derived by SOTIF analysis;

2. The ability of the sensor processing algorithms to accurately model the environment;

3. The ability of the decision algorithms to:

   a. Safely handle functional insufficiencies; and

   b. Make appropriate decisions according to the environment model and the system architecture.

4. The robustness of the system or function;

5. The absence of unreasonable risk of the hazardous behaviour of the intended functionality;

6. The reliability of the assumed infrastructure resources (e.g. sudden outage of communication network or long-term absence of update possibility);

7. The ability of the system (e.g. HMI) to prevent reasonably foreseeable misuse.

10 - Evaluation of known potential hazardous scenarios (Area 2)

To evaluate the identified scenarios as acceptable, the following information can be considered and satisfied:

- Specification and design of the system;

- Identified potential insufficiencies of specification, performance limitations and triggering conditions (including reasonably foreseeable direct misuse) and

- Definition of the verification and validation strategy

11 - Exploration and Evaluation of unknown hazardous scenarios (Area 3)

Unknown scenarios can be encountered in real-life situations. Methods to evaluate the residual risk arising from real-life situations, that could trigger a hazardous behaviour of the system when integrated in the vehicle, are to be analyzed and the analysis shall follow the defined methods listed in the ISO standard.

12 - Criteria for SOTIF release

If operational phase activities have led to SOTIF measures, these measures are reviewed in clause 12. The prerequisite information is reviewed taking the following into account:

1. Are the hazards, functional insufficiencies, and triggering conditions analyzed and any necessary design modifications to achieve the SOTIF implemented and evaluated, to ensure that design modifications have reduced the risk in all specified use cases?

2. Does the verification and validation strategy provide coverage for all the known hazardous scenarios and does it provide an argument for sufficient coverage of unknown scenarios within the scope of the intended functionality?

   a. Does the testing cover identified triggering conditions?

   b. Are sufficient validation activities included in the verification and validation strategy to limit the risk due to known and unknown scenarios?

3. Does the intended functionality achieve a minimal risk condition, when necessary, providing a state without unreasonable risk to the occupants or other road users, considering:

   c. the specified driver intervention;

   d. reasonably foreseeable misuse;

   e. the specified warning to the vehicle occupants and/or the other road users;

   f. the specified degradation of the functionality; and

   g. the DDT fallback (to achieve the minimal risk condition).

4. Is sufficient verification and validation completed and 1495 the validation targets met, to have confidence that the residual risk is not unreasonable?

   h. Has the intended functionality been exercised sufficiently to evaluate both nominal behaviour and potentially hazardous behaviour?

   i. In case of hazardous behaviour, was evidence provided to argue the absence of unreasonable risk?

   j. Did testing provide sufficient coverage argumentation to support the robustness of driving policy across all use cases and/or ODD, OEDR?

5. Are the means necessary to realize operation phase activities (according to Clause 13) operational?

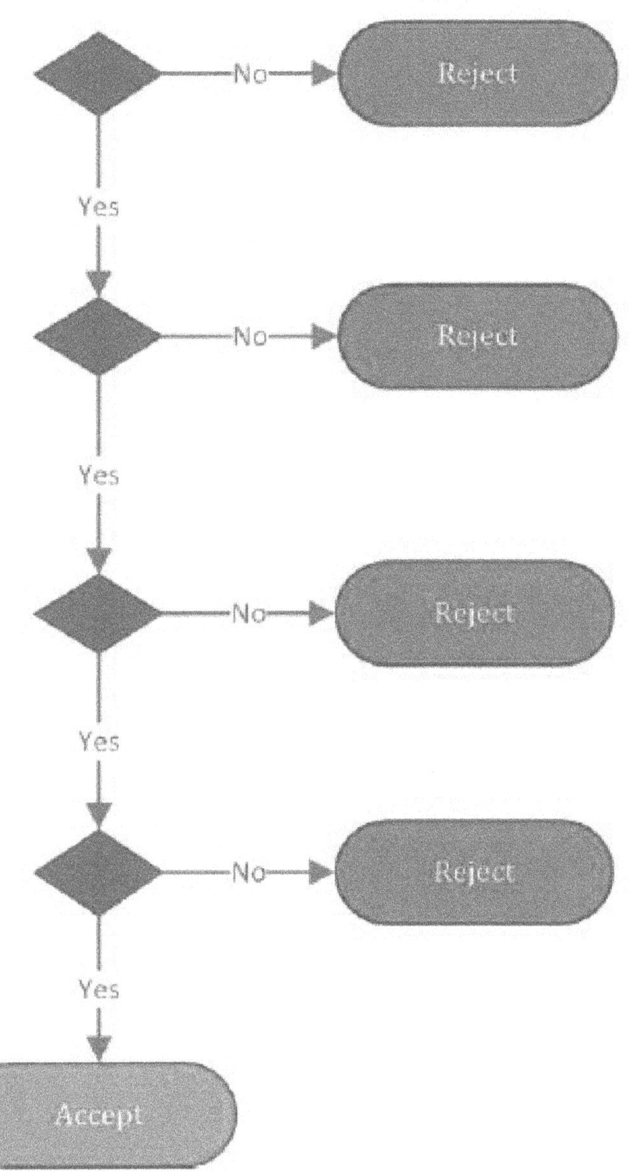

The system is to be released with ISO 21448 SOTIF validated certification only if it satisfies the above listed requirements.

Case Study - SOTIF and ADAS

Advanced Driver Assistance Systems (ADAS) are the hardware, software, and communication systems that are designed to increase safety and reduce risk by using the human-machine interface to aid the humans driving the vehicle. ADAS gives early warnings and reduces the reaction times to potential hazards. This enhanced capability helps prevent injuries and deaths by lowering the number of vehicle accidents, and reducing the serious impact of accidents that cannot be avoided.

SOTIF defines the target and what the intended definition of "functionally safe" is in the situation. The ADAS systems determine how you hit that target by turning intended safety into actual safety in the real world. If engineers don't first define the intended functionality via SOTIF, they have no way of knowing what they are aiming for, and no way of confirming they hit it. And if they don't follow through by deploying validated ADAS solutions, real-world safety is not improved.

In order to make thw world safer we define and apply intended functionality. Applying repeatable scientific principles is more effective than guesswork. All these points underscore the concept of designing with the safety of the intended functionality (SOTIF). And once the SOTIF of the vehicles and roads was defined, engineers began looking for ways to improve the functionality of the drivers themselves.

There are many examples of ADAS systems that have been in use for decades. Some of the more established and less-complex electrical and mechanical systems are anti-lock braking systems, traction control, automatic headlights that turn on in low light conditions, and rearview mirrors that automatically dim. More recent innovations include rearview cameras to help prevent backing accidents, adaptive cruise control that allows you to set and maintain a fixed distance from the car in front of you, navigation assistance in the form of GPS-based graphical and audible systems, hazard avoidance in some or all directions around the vehicle, and lane departure and centering.

The technologies used in these systems can typically be categorized as either those that improve driver awareness and/or those that automate driving tasks. These technologies were selected, and these ADAS systems were designed and built, only after engineers first defined:

- what are the safety problems the systems were intended to solve,

- what a "safe" system looked like, and

- what known and unknown risks might be encountered in their use.

Defining these criteria is the act of defining the safety of the intended functionality.

Lane Keep Assist System

Now if we take an example of Lane keep assist system, the main SOTIF goals would be as follows:

- Lane centering;

- Lane switching, including overtaking or to achieve a minimal risk condition;

- Merge for high and low speed;

- Detect and respond to encroaching vehicles;

- Enhancing conspicuity (e.g., blinkers);

- Detect and respond to no passing zones;

- Detect and respond to lane changes, including unexpected cut-ins;

- Detect and respond to vehicle roadway entry; and

- Detect and respond to relevant adjacent vehicles.

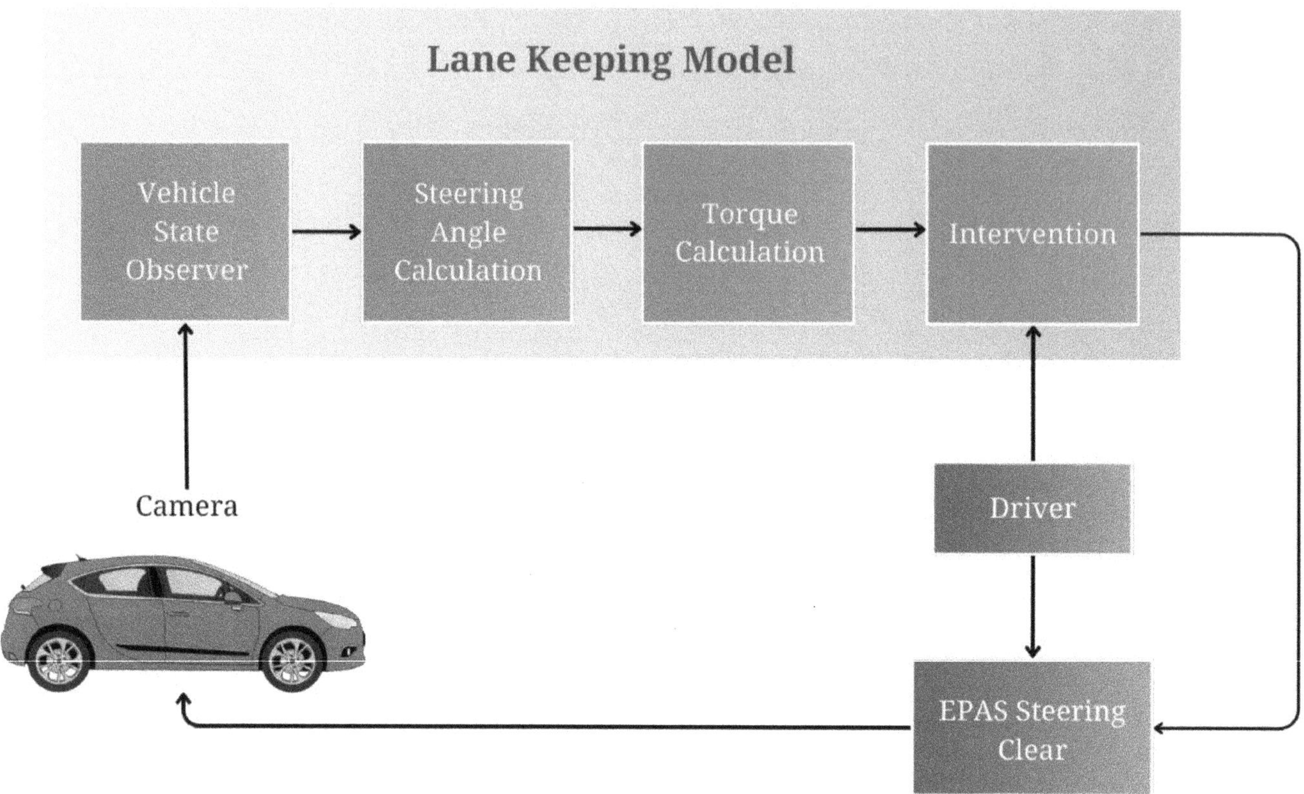

Starting with limitations of the camera module:

Examples of typical camera limitations include:

- Low visibility conditions;

- Low contrast;

- Field of view limits; and

- Missing lane markings or landmarks.

Examples of typical GPS and map limitations include:

- Physical blockage of the satellite signal (e.g., tunnels);

- Complex environments that reflect the satellite signal (e.g., urban canyons);

- Precision; and

- Accuracy of current map data.

Examples of algorithm limitations include:

- Quality and accuracy of the raw sensor data,

- Robustness of the algorithm (e.g., quality and completeness of training data),

- Delays in obtaining sufficient sensor data to update models,

- The theoretical basis of the algorithm (e.g., the probability distribution assumptions, filtering methods),

- Considerations in calculating the confidence intervals and the acceptance limits, and

- Interpretation of complex environments (e.g., multiple vehicles in close proximity).

The sensor perception algorithms considered for proper working of the system include:

- Lane Model – The lane model algorithm is responsible for receiving raw data on the lane markings and road edges from the camera sensor. The lane model algorithm processes this information to determine the lane and roadway boundaries.

- Fusion Tracker – The fusion tracker algorithm receives raw data from the vehicle sensors about objects in a defined boundary surrounding the vehicle. The fusion tracker also receives object classification data from the sensors. The fusion tracker combines the sensor data to create a fusion map that contains the instantaneous position of all detected objects within the defined boundary.

SOTIF - Mandatory for Safe Autonomous Driving

Note that the SOTIF standard originated in a subcommittee from the realm of electronics and general systems, not at the road vehicle level. This makes sense because today's cars are electromechanical and some of the same sensors and devices can be found in other vehicles and applications. The same or similar components might also be shared across on-highway, off-highway, or even aerospace applications.

It is important to understand the added complexity from the beginning. Gone are the days when the car was simple enough to be divided into mechanical and electrical systems. And yet, each application will necessitate requirements tailored to the particulars of its use case and operating environment.

Automotive developers and OEMs put their expertise in a variety of technologies and systems. Although the work on this standard originated in a subcommittee focused on electronics and general systems, that does not affect the priorities of the standard itself. Because each application might utilize a similar component but in different ways to achieve different criteria, it does not make sense for team members to focus first on parts and software. Instead, the standard focuses on, "Safety Of The Intended Functionality," and then provides guidance on the applicable design, as well as verification and validation, measures needed to achieve the SOTIF. This helps to ensure that the expensive design and development work, and the eventual selection of specific technologies, is targeted to an end goal that is clearly defined, rather than engineers plucking possible safety applications out of thin air to justify the use of a given technology.

Chapter 10

Automotive Cybersecurity

1. What is vehicle Cybersecurity?

2. Need for Cybersecurity in Automotive

3. Stages of cybersecurity

4. Threats we face today in Automotive domain

5. Impacts due to Vulnerability

6. Dedicated standards for Cybersecurity

7. Cybersecurity by design

8. Few leading manufacturers in Automotive Cybersecurity

9. Case Study

What is vehicle Cybersecurity?

Vehicle cybersecurity, also known as automotive cybersecurity, refers to the protection of vehicles and their electronic systems from cybersecurity threats and attacks. As modern vehicles become more connected and equipped with advanced technologies, they also become more vulnerable to potential cyber threats. Vehicle cybersecurity aims to safeguard vehicles from unauthorized access, data breaches, manipulation, and any other malicious activities that could compromise the safety, privacy, and functionality of the vehicle.

The automotive industry recognizes the importance of vehicle cybersecurity and is continually working to improve the security of connected vehicles. As vehicles become more technologically advanced, the need for robust cybersecurity measures becomes increasingly critical to ensure the safety and security of drivers, passengers, and other road users.

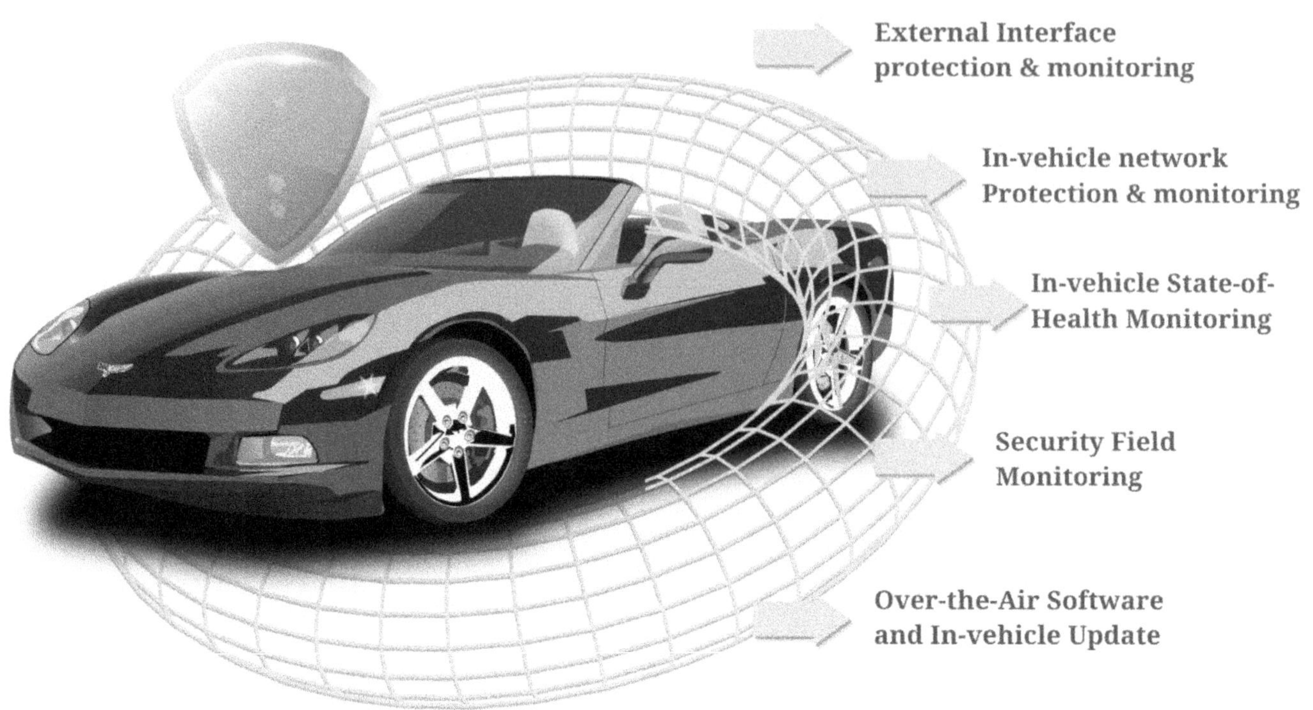

Key aspects of vehicle cybersecurity include:

Connected Car Technology: Many modern vehicles are equipped with advanced features and systems that connect to the internet or other external networks. These features can include infotainment systems, GPS navigation, cellular connectivity, and telematics. Securing these connections is crucial to prevent unauthorized access and potential hacking attempts.

Electronic Control Units (ECUs): Modern vehicles have numerous electronic control units that control various aspects of the vehicle's operation, such as engine control, braking, and steering. Securing these ECUs is essential to prevent unauthorized manipulation that could compromise the vehicle's safety.

Data Protection: Vehicles collect and process a significant amount of data, including personal information about drivers and passengers. Ensuring the privacy and protection of this data is vital to prevent data breaches and potential misuse.

Authentication and Access Control: Implementing robust authentication mechanisms and access control protocols helps ensure that only authorized personnel can access sensitive vehicle systems and data.

Over-the-Air (OTA) Updates: OTA software updates are increasingly common in modern vehicles to provide bug fixes and new features. Securing OTA updates is critical to prevent potential hacking attempts through update channels.

Intrusion Detection and Prevention: Employing advanced intrusion detection and prevention systems helps identify and block potential cyber threats in real-time.

Regulatory Compliance: Vehicle cybersecurity must comply with relevant industry and government regulations to ensure that manufacturers meet specific security standards.

Collaboration with Manufacturers: Vehicle manufacturers work closely with cybersecurity experts to identify vulnerabilities and improve the security of their vehicles.

Continuous Monitoring and Response: Regular monitoring and timely response to potential cybersecurity incidents are essential to minimize risks and mitigate the impact of any attacks.

The concept of vehicle cybersecurity emerged as cars became more computerized and connected to external networks. The timeline for the existence of vehicle cybersecurity can be traced back to several key milestones. The first instances of vehicle cybersecurity issues were reported during Late 1990s - Early 2000s, mainly related to electronic control units (ECUs) and onboard diagnostics. Researchers and hackers demonstrated vulnerabilities in car systems, raising concerns about potential cyber threats to vehicles.

In 2002, A team of researchers from the University of Washington and the University of California published a paper titled "Experimental Security Analysis of a Modern Automobile," which highlighted the security weaknesses in automotive systems and the potential for remote exploitation.

In 2007, Researchers at the University of Washington and the University of California demonstrated a proof-of-concept attack on a connected car, remotely taking control of its functions, including braking and acceleration. In 2010s, vehicles became more connected and equipped with advanced infotainment systems, Bluetooth, Wi-Fi, and cellular connectivity, the need for robust cybersecurity measures became more apparent. Various automakers started focusing on improving security and encryption in their vehicles. In 2015, The Jeep Cherokee hack, where security researchers remotely accessed and controlled a Jeep's systems, garnered significant media attention. This incident raised public awareness about the importance of vehicle cybersecurity. In 2016, The National Highway Traffic Safety Administration (NHTSA) and the automotive industry established the Automotive Information Sharing and Analysis Center (Auto-ISAC) to enhance collaboration and information sharing on cybersecurity threats. Several cybersecurity companies started offering specialized automotive cybersecurity solutions and services to automakers, aiming to identify vulnerabilities and provide solutions to enhance the security of connected vehicles.

The U.S. Department of Transportation released the "Automated Vehicles 3.0" guidance, which included a section on cybersecurity best practices for connected and automated vehicles. In 2020, The automotive industry continues to invest in research, development, and implementation of advanced cybersecurity measures to protect connected vehicles from potential cyber threats.

Over the past two decades, vehicle cybersecurity has evolved significantly, becoming an integral part of the automotive industry's focus on safety and consumer trust. As connected and autonomous technologies continue to advance, the importance of vehicle cybersecurity will only continue to grow to ensure the safety and security of future smart vehicles.

Need for Cybersecurity in Automotive

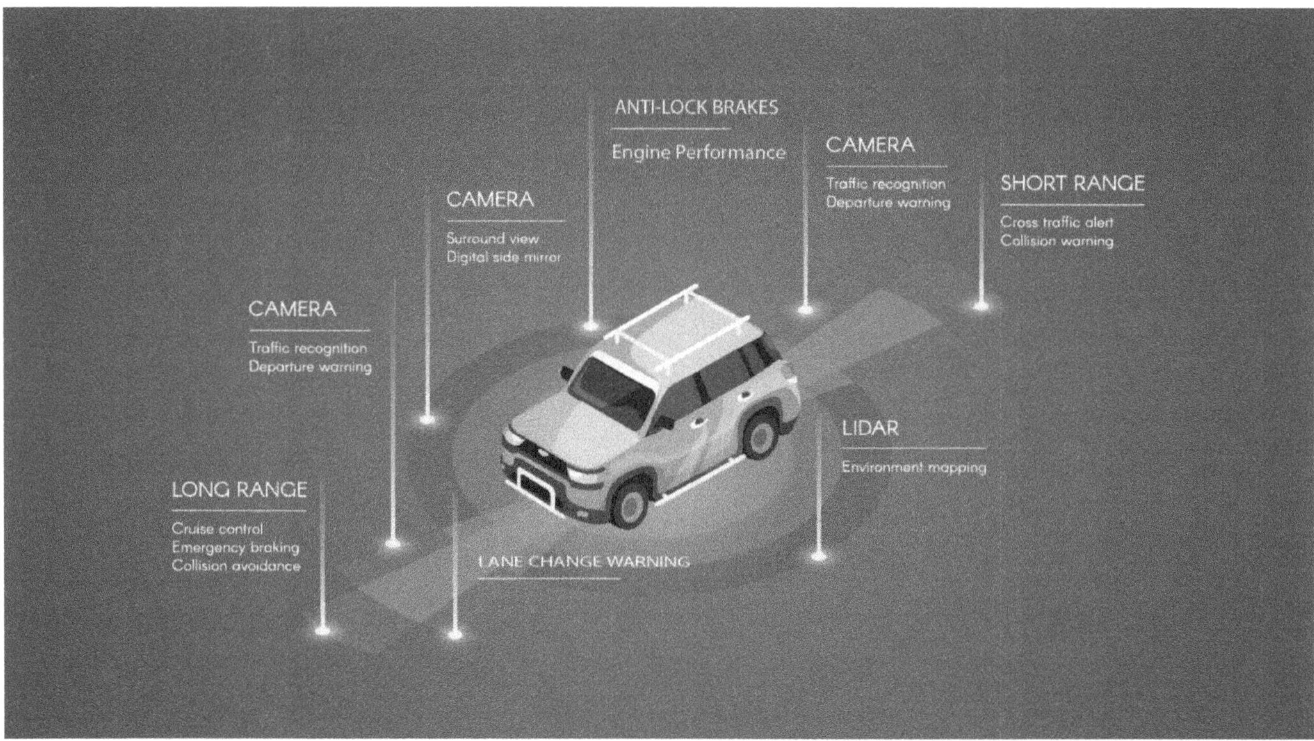

The need for cybersecurity in the automotive industry has grown significantly due to the increasing adoption of connected technologies and advanced features in modern vehicles. As vehicles become more like computers on wheels, they face cybersecurity risks similar to those encountered in other connected devices. Here are some key reasons why cybersecurity is essential in the automotive sector:

Vehicle Safety: Cybersecurity is crucial for ensuring the safety of drivers, passengers, and other road users. With the integration of advanced driver assistance systems (ADAS) and autonomous driving technologies, any cyber-attack on vehicle systems could potentially lead to loss of control, accidents, or even fatalities.

Protection against Hacking: As vehicles become more connected, they are susceptible to hacking attempts. Unauthorized access to a vehicle's electronic control systems can lead to dangerous situations, such as remote control of critical functions like braking or steering.

Data Privacy: Modern vehicles collect and store large amounts of data, including personal information about drivers and passengers. Ensuring data privacy is essential to protect individuals' sensitive information from falling into the wrong hands.

Preventing Theft and Tampering: Cybersecurity measures can help prevent vehicle theft and tampering, as hackers may attempt to bypass security systems to gain unauthorized access to the vehicle.

Preventing Fraud: Connected vehicles are vulnerable to identity theft and fraudulent activities. Cybersecurity helps protect against fraudulent access to vehicle services and accounts.

Securing OTA Updates: Over-the-Air (OTA) updates are common in modern vehicles to enhance functionality and address software vulnerabilities. Proper cybersecurity ensures the integrity and security of OTA updates, preventing unauthorized modifications or malicious software injection.

Maintaining Consumer Trust: Cybersecurity is essential to build and maintain consumer trust in connected vehicles. If consumers feel that their safety and privacy are at risk, they may be hesitant to adopt new technologies and features.

Compliance and Regulatory Requirements: The automotive industry is subject to various regulations and standards concerning cybersecurity. Compliance with these requirements is necessary to meet industry standards and ensure customer safety.

Brand Reputation: A cybersecurity breach in the automotive sector can severely damage a manufacturer's reputation and credibility, leading to financial losses and loss of customer loyalty.

Evolving Threat Landscape: The threat landscape is continually evolving, and cyber-attacks are becoming more sophisticated. Continuous cybersecurity efforts are necessary to stay ahead of potential threats.

Given these factors, automotive manufacturers and stakeholders must prioritize cybersecurity throughout the vehicle development lifecycle. Collaborating with cybersecurity experts, implementing robust security measures, and staying up-to-date with the latest threats are essential to protect connected vehicles and the safety of their users.

Stages of cybersecurity

Indeed, important assets need to be protected using a multilayered security approach in order to reduce the impact of a successful intrusion. Similarly, it is certainly a good practice to use multiple security countermeasures to mitigate risks in connected cars, in case an intruder is able to get access through several breaches.

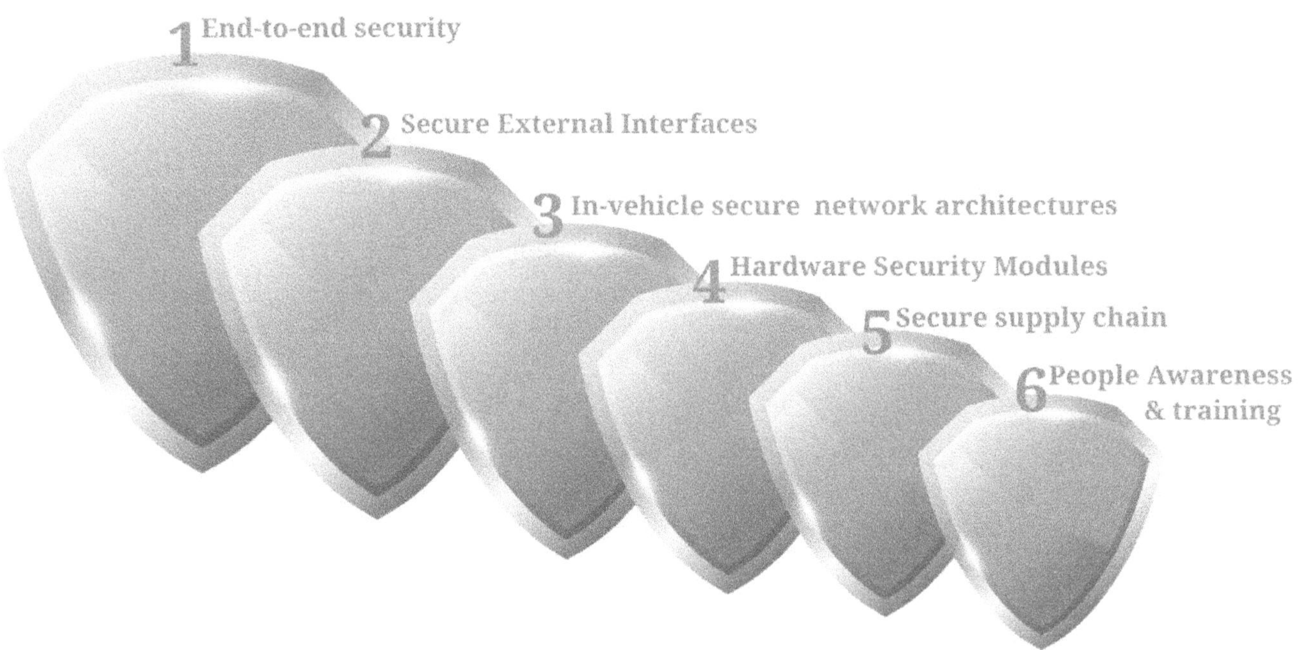

A security framework for automotive should be then built upon a defense in depth strategy including:

Secure interfaces with the external world (e.g. OTA, driver dedicated applications, OBD, Bluetooth). Indeed, these interfaces can be seen as an explicit invitation to hack a vehicle system as several exploits already exist on them. They should be major points of interest and a priority for automotive architects when the security policy is being designed,

In-vehicle secure network architecture providing physical segregation and isolation of safety related ECUs using secure gateways and communication buses (e.g. CAN, Ethernet, FlexRay),

Hardware Security Modules (HSM) which provide a strong security anchor for software by protecting basic security functions (e.g. secure boot, key generation, key storage, active memory protection) for microcontrollers. HSM help to deliver hardware security services such as trusted Execution Environment (TEE) or cryptographic computing acceleration for better performances,

Secure the supply chain as several actors are usually involved in connected car system design. All the stakeholders need to be aware of the cybersecurity risks to be mitigated and act accordingly by following guidelines and best practices they are concerned with,

End-to-end security strategy by protecting the chain-of-trust from the car architecture to the servers and the cloud.

Threats we face today in Automotive domain

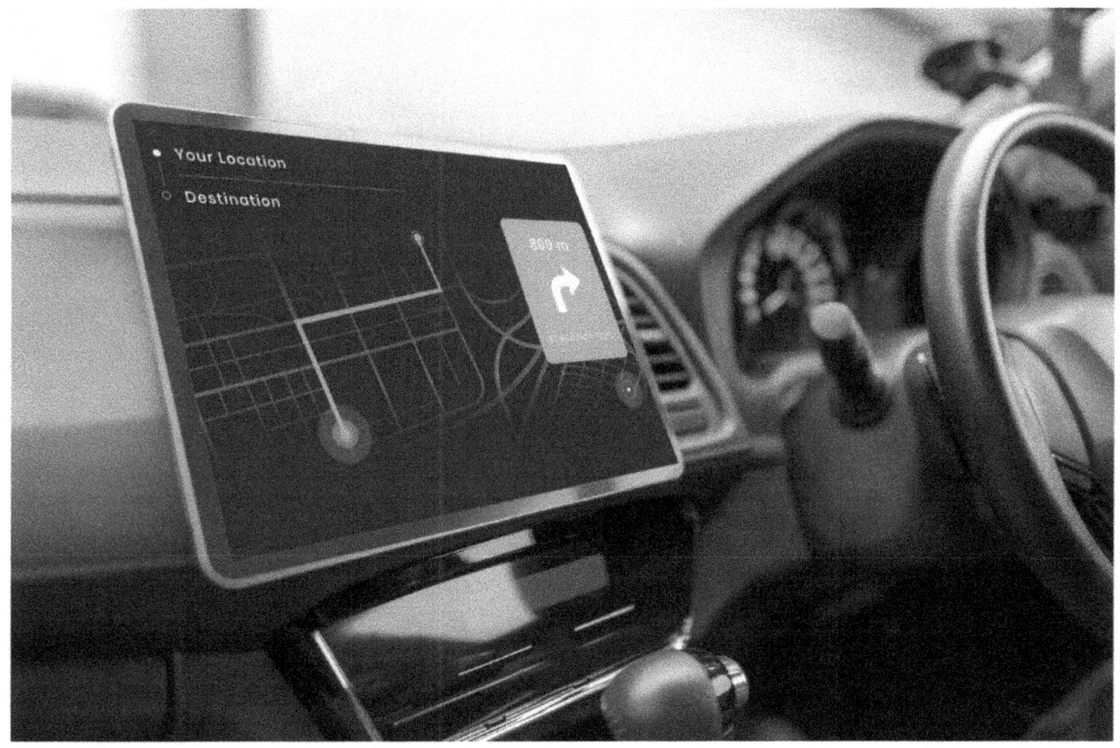

The complexity of car's electronic systems is continuously growing and driven by the acceleration of the market requirements. The market expects the car to be safe, not only by protecting its occupants but also by preventing accidents. This is achieved by electronic driver assistance systems (e.g. parking, speed regulation, lane, and blind spot detection, pre-collision). The car is also expected to be more comfortable, hence the comfort systems (e.g. automatic cooling, seat adjustment with memory, automatic tailgate opening, and performance control) and to provide a complete infotainment system (e.g. navigation, audio, voice assistant, Bluetooth). Car manufacturers are providing more added value services requiring network connections (e.g. emergency calls, remote diagnostic, remote support, internet browsers, and concierge).

Modern cars have to provide to their users a continuum in their lives. People want to be permanently connected to get various information and also to interact with their social and professional ecosystems. These requirements lead to automotive electronic architectures which are growing rapidly in complexity and which are large even in comparison to other industries. The average modern high end car software is 100 million lines of code, to be compared with Windows 7 (39.5 million in 2009) or a Boeing 787 (13.8 million). Having so many lines of codes implies that some vulnerabilities very likely exist and represent security issues.

Amount of connected cars is growing fast. In a global market growing by more than 70 million cars per year, the amount of connected cars will grow fast. This will significantly increase the attack surface hackers might exploit.

The reasons why they are vulnerabilities in software are diverse. It appears that the management of large automotive projects is not always aware enough of the importance and specificities of cybersecurity. One of them is linked to the fact that developers of software in the automotive sectors are not used to taking into account security since the beginning of the project. They consider they are not enough trained and that there is not enough appropriate security enabling technologies in the processes they use. Security awareness, and active management of security policy at high levels in an organization, is also typically a challenge for industries facing new cybersecurity threats.

Legal regulation must be reinforced to establish a collective pressure to improve protection. Another lever that emphasizes security in automotive is the evolution of the regulations. To take one single example, the European General Data Protection Regulation (GDPR) was made effective in May 2018. Mastering data privacy in vehicles will be a challenge considering the rising amount of data stored and managed in a vehicle and between the vehicle and ground bases. The challenge will even be higher for rented cars, fleets and car sharing services. Indeed, privacy related data might be stored and shared for technical or commercial purposes. The customer must be formally informed and must confirm his consent. This might become complex when services are delivered by third parties and related gathered data sold for added value services.

Modern society cannot accept anymore that security incidents impacting end users be kept hidden. There is a global trend of the regulators to oblige organizations to make security incidents publicly known. One of the reasons is that making information public will lead to establishing a collective pressure to improve protection. Without this pressure, who knows if the improvement would be performed?

Regulation requires that some repair and maintenance information (RMI) has to be easily and clearly accessible to promote competition in the vehicle repair market. This limits some of the measures available for managing and controlling security.

Lack of commonly accepted standards for development and accreditation In order to successfully address security in a connected world, all stakeholders should share a common understanding of how to manage security: what is security, how to implement it, how to control it, what are the processes and organization necessary to manage it? In diverse industries, information security management and implementation is described in standardized best practices documents. There is not yet such a widely recognized set of reference documents for the automotive industry. Some best practices and standards exist but are not covering the full life cycle of security management and are not yet widely recognized by all industry players. This generates various side effects at industry level, such as engineering complexity, maintenance and integration issues, complexity of monitoring and investigating to name a few. This requirement needs to be a top priority for car manufacturers, suppliers, and stakeholders urgently.

Impacts on Safety due to Cyber Vulnerability

Having connected the car to the external world created new risks. Systems that were designed for years as being completely disconnected and exchanging information exclusively within the car– when on the move - have been now connected to billions of devices, computers and objects across the world. Despite the fact that the security of some of these connections may be strengthened by car makers or OEMs, the situation creates new risks. What if the embedded systems driving the brakes, assuring geo positioning or ensuring lane following would be corrupted? This might lead to an accident with consequences potentially impacting persons transported by the vehicle, but also people and goods located in the surrounding environment. Therefore the connection of the car leads to drastically influence safety. What might be seen as a security issue – illegitimately accessing and modifying data in vehicles – is now a safety issue. Security and safety must converge for our best interest.

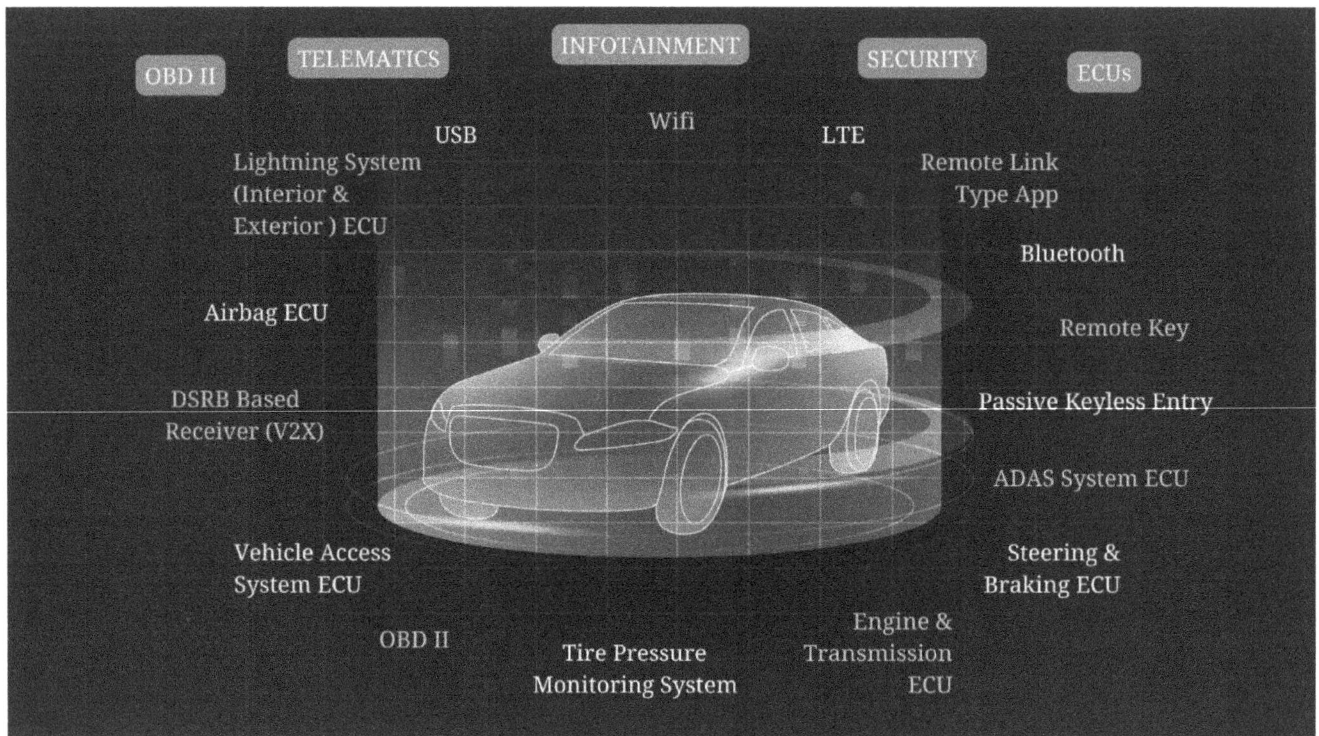

The increasing integration of technology in vehicles has brought numerous benefits, such as improved connectivity, automation, and convenience. However, it has also introduced new risks and challenges, including cyber vulnerabilities that can impact vehicle safety. Here are some potential impacts of cyber vulnerabilities on vehicle safety:

- Unauthorized Access and Control: If a vehicle's electronic control systems are compromised, hackers could gain unauthorized access to critical functions, such as braking, acceleration, and steering. This could lead to loss of control over the vehicle and potentially cause accidents.

- Remote Exploitation: Cyber vulnerabilities could allow attackers to gain remote access to a vehicle's systems, bypassing traditional security measures. This means that a vehicle could be controlled or manipulated from a remote location, posing serious safety risks to the occupants and others on the road.

- Malware and Ransomware: Vehicles are becoming more like computers on wheels, and just like any other computer, they are susceptible to malware and ransomware attacks. Malicious software could disrupt normal vehicle operations, leading to safety hazards or rendering certain safety-critical functions inoperative.

- Data Privacy Concerns: Modern vehicles collect and store a vast amount of data related to driver behavior, location, and personal preferences. If cyber vulnerabilities expose this data to unauthorized individuals, it could lead to privacy breaches, stalking, or other malicious activities that could affect the safety of vehicle occupants.

- Over-the-Air (OTA) Updates Risks: Automakers often provide OTA updates to enhance vehicle functionality and address security issues. However, if these updates are not adequately secured, hackers might exploit the update process to introduce malicious software into the vehicle's systems.

- Supply Chain Vulnerabilities: Vehicles are composed of various components and subsystems provided by different suppliers. Cyber vulnerabilities in any of these components could potentially affect the overall safety and performance of the vehicle.

- Distracted Driving: Infotainment systems and other connected technologies in vehicles can be vulnerable to hacking, leading to potential distractions for the driver. Distractions while driving can significantly increase the risk of accidents.

- Delayed or Ineffective Recalls: If a cyber vulnerability is discovered in a vehicle model after it has been released to the market, it may require a recall to address the issue. Delays or difficulties in implementing effective recalls could leave vehicles exposed to potential cyber-attacks for extended periods, compromising safety.

In modern cars, both safety and data security are critical; a hacker can threaten your life by hacking the car or use stolen personal data to commit serious fraud. To mitigate the impacts of cyber vulnerabilities on vehicle safety, automakers, government agencies, and cybersecurity experts are taking various measures, including:

Regular Security Updates: Automakers must continuously update and patch software systems to address known vulnerabilities and respond to emerging threats.

Enhanced Cybersecurity Testing: Robust cybersecurity testing during the development and production phases can help identify and address vulnerabilities before vehicles reach the market.

Collaboration and Information Sharing: Sharing information about cyber threats and vulnerabilities among industry stakeholders can improve collective defense against attacks.

Secure Communication: Implementing secure communication protocols and encryption methods can protect sensitive data transmitted between the vehicle and external networks.

Hardware and Software Redundancy: Incorporating redundancy in critical vehicle systems can provide fallback options in case of cyber-attacks or system failures.

Public Awareness and Education: Educating vehicle owners about the potential risks and best practices for ensuring cybersecurity can improve overall safety.

Government Regulations: Establishing and enforcing cybersecurity standards for the automotive industry can set a baseline for safety and security practices.

As technology continues to advance, addressing cybersecurity in vehicles will remain a crucial aspect of ensuring safe transportation for all road users.

Dedicated standards for Cybersecurity

A well-established standard is always a guarantee that processes and implementations are compliant with best practices and guidelines. Several efforts have already been undertaken in order to provide such cybersecurity guidelines for the automotive industry, globally but also country-specific. There is a risk of overlapping and all these standardization bodies need to coordinate with each other to avoid conflicts and ambiguities. Local initiatives include for instance the EVITA (E-Safety Vehicle Intrusion Protected Applications) project in Europe which was aimed to provide in-vehicle reference architecture based on HSM. The Japanese IPA (Information Promotion Agency) vehicle information security guide covered an end-to-end life-cycle of the vehicle including third-party and suppliers behavior toward security. More recently, international standardization bodies such as ISO (International Organization for Standardization) and SAE (Society of Automotive Engineers) joined their effort to work on the definition of a dedicated cybersecurity standard for the automotive industry. Works have already initiated internally within the Vehicle Electrical System Security Committee where the J3061 cybersecurity guidebook and the J3101 requirements for hardware-protected security documents are produced. ISO's TC22 and SAE are also identifying the potential interactions between system safety and cybersecurity.

Establishing a shared terminology and unique standard that can be universally used by stakeholders, third party suppliers, developers and car manufacturers, it helps in raising the overall awareness around the topic and inculcates a 'built-in' security culture within the industry. Altran also shares the interest for working together with other security standardization activities in other industries such as intelligent transport systems or IOT. Lessons learned from similar past experiences are always profitable and useful. Based on our experience in other similar fields and standardization activities, we look forward to actively debating the core components and work initiated in the SAE J3061 initiative.

Below are few initiatives to form a common cybersecurity standard for automotive industry

Body	Working Group	Objectives
SAE	Vehicle Electrical System Security Committee	Provide a Cybersecurity Guidebook for Cyber-Physical Vehicle Systems. Requirements for Hardware-Protected Security for Ground Vehicle Applications.
ISO	TC22 /SC 32/ WG11	Coordinate standardization activities with SAE Vehicle Electrical System Security Committee for Automotive security engineering.
ETSI	TC ITS WG5	Assurance of ITS solutions conformity to regulatory requirements for privacy, data protection, lawful interception and data retention.
IEEE	SCC42/SCC Type 2	Coordination of IEEE standardization activities for technologies related to transportation, especially in the areas of connected vehicles, autonomous/automated vehicles, inter- and intra-vehicle communications, and other types of transportation electrification.

Cybersecurity by design

One of the common and effective solutions discussed broadly is to ensure cybersecurity by design.

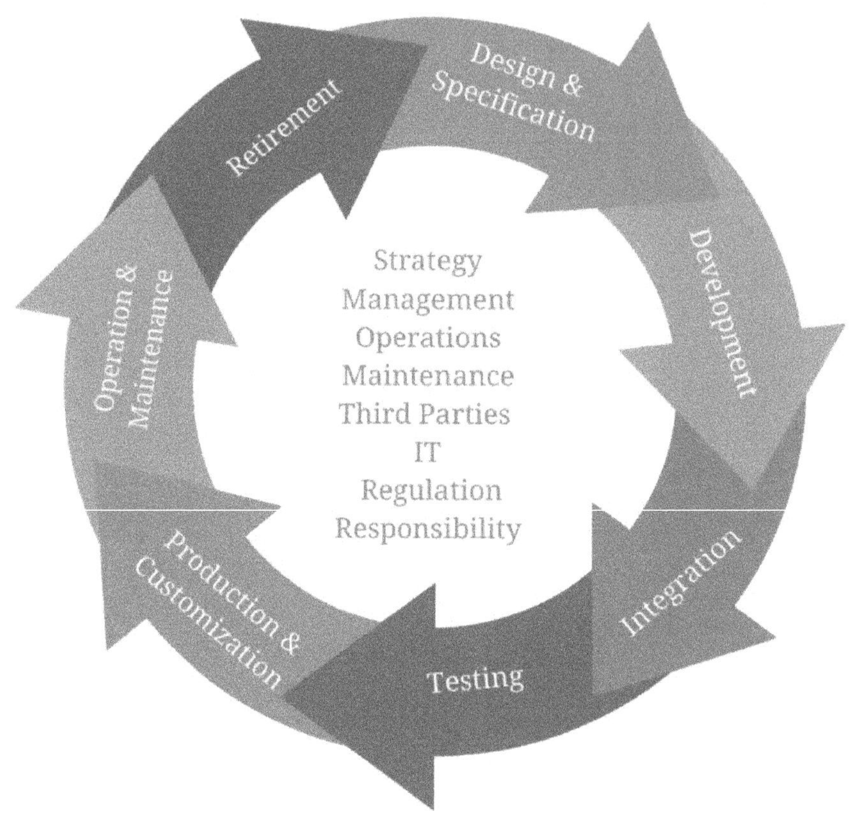

Security By Design Across car development lifecycle

Security is taken into account in every step of the project lifecycle starting from specifications to validation. As an example, secure coding rules allow developers to rely on strong security practices while their code is produced, avoiding inherent vulnerabilities such as buffer overflow. A global top-down security strategy needs to be defined to fulfill such an objective, covering management, operations, maintenance, third parties, car manufacturers, and suppliers. Security by design implies that a 'built-in' security is implemented at early stages of the design of the car system, compared to a 'built-on' security where security is added block by block to counterfeit newly discovered security breaches.

A good security-by-design policy should then cover

- Organizational methods for secure development, by conducting for instance repeated risk assessments on most critical components in the car system,

- Secure management of related projects, providing security requirements and technical solutions,

- Security verification and validation, existing security development lifecycle frameworks such as ISO/IE27034 can be reused and adapted to achieve security-by-design in automotive systems (e.g. penetration testing, static code analysis).

Example Strategies for Cyber Attack countermeasures

In order to reduce the attack surface and protect most critical assets in a car system against the variety of threats discussed above, several effective state-of-the-art security countermeasures can be applied such as:

- Secure in-vehicle communications using mature COTS cryptographic products for primary functions such as onboard network segregation, intrusion detection, or data filtering,

- ECUs hardening by respecting best practices such as deleting interfaces and services used for development when the car is ready for release, or using a tradeoff combination between hardware and software security to leverage defense-in-depth. These practices should be applied to every ECU, even those which are not critical to the vehicle operation,

- Perform regular survey on cybersecurity evolution (both from an intrusion and defense point of view) in similar intelligent transportation systems such as Aeronautics where the embedded context is comparable to in-vehicle networks (e.g. partitioned operation systems, domain segregation, transport operation criticality),

- Rely on technology diversity to counter security monoculture. The idea is to do not put all eggs in the same basket: diversity, when properly applied, is a practical tool to avoid attack propagation to critical components,

- Maintain security over time using surveillance techniques such as continuous vulnerability management or firmware and software updates using out-of-band channels. Also, security maintainability is a key enabler for long-term cryptographic-based protection. It is unlikely to have a cryptographic key strong enough to last for vehicle lifetime; these systems should be then thought of with such constraints in mind.

Few leading manufacturers in Automotive Cybersecurity

APTIV

Aptiv develops software and computing platforms for self-driving vehicles. The company's cybersecurity tools protect everything from a car's infotainment system to its wiring

Aptiv's self-driving cars were the first to be tested in Singapore and are slated to be fully implemented there by 2022. In collaboration with the ride-hailing service Lyft, Aptiv has already provided more than 100,000 robo taxi rides to passengers in Las Vegas and plans to expand its service within the next two years.

NVIDIA

NVIDIA uses AI-powered data processors and chips to operate and protect self-driving cars. The company's software and cloud-based technologies help autonomous vehicles securely learn and relay driving data

The NVIDIA deep learning systems have been used by Tesla, Mercedes-Benz, Audi, Toyota, Volkswagen and more to power and protect self-driving vehicles.

Centri Technology

Centri makes cybersecurity solutions for IoT-enabled devices in autotech. The Centri IoTAS installs on chips and mobile apps to protect automobile sensors as well as the data that helps cars learn important driver navigation preferences like optimal routes and addresses.

The IoTAS platform requires no internet connection to protect IoT-enabled autotech devices. Instead, it connects to all trusted devices with identity management technology that requires around 40 percent less bandwidth than BLE and TLS.

Dellfer

Dellfer is an automotive cybersecurity startup focusing on coding for autotech software. The company's embedded code helps IoT-enabled cars battle cyber attacks throughout a car's system. No Internet connection is needed to update critical patches. Instead, the company deploys code execution paths at runtime for security enforcement.

Dellfer partnered with DENSO, the world's second-largest mobility provider, to help install Dellfer's IoT cybersecurity tools in a wider range of vehicles.

Argus CyberSecurity

Argus provides commercial smart vehicles with anti-cyber attack tools like connectivity and in-car network protection that safeguard everything from a vehicle's infotainment center to the communication networks that run between its software and hardware.

Continental, the automotive parts manufacturing company, now integrates Argus' cybersecurity solutions into all of its connected vehicle electronics.

GuardKnox

GuardKnox creates coding architecture for autonomous cars that operates everything from the general vehicle systems (including sensors) to tools that enhance a car's user experience (infotainment systems, center consoles and more).

Porsche enlisted GuardKnox to improve cybersecurity in its new line of vehicles. The German carmaker says the new technology will protect against hacking attacks and act as a foundation for "real-time customization of the vehicle."

Harman

Harman partnered with IBM to develop the Harman SHIELD, which protects key entry points of a car's network from hackers. In addition, it continuously performs a threat analysis to determine which points are most vulnerable at any given moment.

Harman revealed the Ignite 3.0, an automotive-based assistant for cars, at the 2019 CES conference in Las Vegas. The technology is backed by the company's SHIELD cybersecurity infrastructure.

Intertrust

Intertrust makes products that help personalize drivers' cybersecurity needs and overall experience. Some of the company's products include tools that protect a vehicle's infotainment center, prevent unauthorized entry and stop the gathering of personal data.

One of Intertrust's main software products, whiteCryption, speeds up and safeguards content delivery to drivers. Another tool, Personagraph, encrypts a driver's personal data.

Thales

Thales provides electrical system engineering services for IoT hardware, software, devices and vehicles. Its services within the transportation and automotive industries operate with an emphasis on cybersecurity and data regulation compliance.

In 2021, Thales announced a partnership with Gireve, a company created to further connect electric mobility operators in Europe. Collaboratively creating the "Plug & Charge" system, this project allows drivers to charge their electric vehicle at any participating compatible charging station, and be automatically billed through a secure card-free transaction.

Case Study - Hacking Jeep Cherokee: Unraveling the Security Vulnerability

Introduction:

In 2015, two cybersecurity researchers, Charlie Miller and Chris Valasek, made headlines by demonstrating how they successfully hacked into a Jeep Cherokee's computer system, remotely taking control of critical functions. Their groundbreaking experiment shed light on the potential dangers of cyber vulnerabilities in modern vehicles and raised significant concerns among automakers and consumers alike.

Background:

Charlie Miller and Chris Valasek were well-known in the cybersecurity community for their expertise in identifying weaknesses in vehicle software and exploiting them for research purposes. They decided to focus their efforts on the Jeep Cherokee, a popular and technologically advanced SUV manufactured by Fiat Chrysler Automobiles (FCA).

The Vulnerability:

The duo discovered a security vulnerability that allowed them to gain unauthorized access to the Jeep Cherokee's infotainment system, known as the Uconnect. This system is responsible for controlling various non-critical functions, such as navigation, entertainment, and internet connectivity. Through the Uconnect system, Miller and Valasek were able to pivot into the vehicle's more critical systems, including those related to braking, acceleration, and steering.

Exploiting the Vulnerability:

To exploit the vulnerability, the hackers used a laptop connected via a wired connection to the Jeep Cherokee's diagnostic port. Through this connection, they managed to infiltrate the Uconnect system and upload their custom-made software. Once inside, they were able to manipulate various functions remotely, demonstrating their control over the vehicle.

During their experiments, they showed the following alarming capabilities:

Taking Control of Steering: The hackers remotely disabled the vehicle's steering, demonstrating how they could alter its direction while the driver had limited control.

Braking Manipulation: Miller and Valasek remotely interfered with the Jeep's braking system, reducing its effectiveness and potentially causing dangerous situations.

Acceleration Control: They were able to control the acceleration of the Jeep, potentially causing sudden speed changes without driver input.

Shutting Down the Engine: The researchers demonstrated the ability to remotely shut down the engine while the vehicle was in motion.

Impact and Response:

The live demonstration of these exploits caused widespread concern about the cybersecurity of modern vehicles. FCA issued a voluntary recall of approximately 1.4 million vehicles, including Jeep Cherokees, to address the security vulnerabilities discovered by Miller and Valasek. The recall aimed to provide a software update to patch the vulnerabilities and strengthen the Uconnect system's security.

Furthermore, the incident prompted a broader industry-wide focus on automotive cybersecurity. It highlighted the need for robust security measures, regular software updates, and the establishment of bug bounty programs that incentivize ethical hackers to report vulnerabilities to manufacturers.

Conclusion:

The hacking of the Jeep Cherokee by Charlie Miller and Chris Valasek served as a wake-up call for the automotive industry, signaling the critical importance of addressing cybersecurity vulnerabilities in modern vehicles. Their research led to significant improvements in vehicle security measures and continues to shape the ongoing efforts to ensure the safety and privacy of drivers and passengers in an increasingly connected automotive landscape.

In 2016, the National Highway Traffic Safety Association (NHTSA) published Cybersecurity Best Practices for Modern Vehicles, a brief guide offering non-binding guidance on cybersecurity for the auto industry.

NHTSA's Response - Layered approach to cybersecurity

In response to the hacking of the Jeep Cherokee and the potential risks it exposed in terms of vehicle cybersecurity, the National Highway Traffic Safety Administration (NHTSA) took several actions to address the situation and enhance automotive cybersecurity standards. The NHTSA is the federal agency responsible for regulating vehicle safety and has a crucial role in ensuring the protection of consumers from emerging threats, including cyber vulnerabilities in vehicles.

NHTSA Investigation: Following the publicized demonstration of the Jeep Cherokee hack, the NHTSA launched an investigation into the security vulnerability and its potential implications for vehicle safety. They collaborated with Fiat Chrysler Automobiles (FCA) and other relevant stakeholders to understand the extent of the issue and assess the potential risks to drivers and passengers.

Issuing a Recall: As a result of the investigation, the NHTSA played a pivotal role in overseeing FCA's recall of approximately 1.4 million vehicles, including certain models of the Jeep Cherokee. The recall aimed to address the security vulnerabilities identified by the researchers and ensure that affected vehicles received necessary software updates to enhance cybersecurity.

Security Guidance and Regulations: In response to the growing concerns about automotive cybersecurity, the NHTSA started working on developing security guidance and regulations for automakers. The agency recognized the need to set standards and best practices to ensure that manufacturers implement robust security measures to protect vehicles from potential cyber threats.

Collaboration with Industry and Researchers: The NHTSA actively engaged with the automotive industry, cybersecurity experts, and researchers to understand evolving threats and identify effective solutions. They encouraged collaboration and information sharing to address cybersecurity challenges collectively.

Awareness and Education: The NHTSA played a role in raising public awareness about the importance of automotive cybersecurity. They communicated safety recommendations to consumers and encouraged vehicle owners to stay informed about software updates from manufacturers.

Encouraging Bug Bounty Programs: To incentivize ethical hacking and responsible disclosure of vulnerabilities, the NHTSA encouraged automakers to establish bug bounty programs. These programs offer rewards to security researchers who responsibly report identified vulnerabilities to manufacturers, allowing them to be addressed promptly.

Continued Monitoring and Adaptation: The NHTSA recognized that cybersecurity threats would continue to evolve, and the automotive industry needed to stay vigilant. They committed to monitoring the landscape closely and updating their regulations and guidance as necessary to address emerging challenges.

Overall, the NHTSA's response to the hacking of the Jeep Cherokee demonstrated its commitment to ensuring the safety and security of vehicles on the road. By taking proactive measures to investigate the issue, oversee recalls, develop guidelines, and collaborate with stakeholders, the NHTSA aimed to strengthen automotive cybersecurity practices and protect consumers from potential cyber threats in the future.

Chapter 11

Functional Safety of Off-road vehicles

Topics to be covered

1. Need of Functional Safety in all machinery

2. Industrial Standards implementing Functional Safety

1. FuSa Standards of Off-highway Machinery

2. Terms used for FuSa HW and SW

3. Common Functional Safety Lifecycle

4. Determination of AgPLr and MPLr

5. Safety Concepts

6. Safety requirements at various project phases

7. Validation and verification of safety requirements

8. Assessment of Functional Safety

Need of Functional Safety in all machinery

Functional Safety is a process of achieving the absence of unreasonable risk caused by hazards due to system malfunctions. The functionally safe system detects the potentially dangerous condition and responds with corrective action. It includes the safety measures that minimize risks by adding monitoring and protective functionality.

Safety related applications like Braking, Steering etc. are present in all domains wherever machinery is used. Let it be airways, railways or robotics in factories, every machinery possesses a potential risk and the functional safety standards derived from IEC61508 for the particular domain aims to reduce that potential high risk to a reasonable risk so the machinery of that domain can be accepted to be around life and valuable properties.

Let us take an example of a tractor in an agricultural field, failure to apply brakes can run over bystanders or collide with another machine/vehicle and an uncommanded steering can collide with vehicles or pedestrians.

Industrial Standards implementing Functional Safety

Now that we discussed how functional safety is mandatory to be implemented for all the fields involving machinery, let us see some of the standards derived from the master IEC61508 for defining safety requirements for specific domains.

Some of the are different fields that are expected to implement functional safety are:

- Automotive

- Railways

- Aerospace

- Medical devices

- Agricultural machines

- Construction machines

- Power Plants

IEC 61508 is a basic functional safety standard applicable to all industries. It defines functional safety as: "part of the overall safety relating to the EUC (Equipment Under Control) and the EUC control system which depends on the correct functioning of the E/E/PE safety-related systems, other technology safety-related systems and external risk reduction facilities." The fundamental concept is that any safety-related system must work correctly or fail in a predictable and safe way.

Functional safety, as per IEC61508, is achieved when all the specified safety functions, of the safety-related system, can satisfy their required safety performance. Functional safety is undertaken by active systems.

IEC 61508 sets out the requirements for ensuring that systems are designed, implemented, operated and maintained to provide the required safety integrity level (SIL). Four SILs are defined according to the risks involved in the system application, with SIL4 being used to protect against the highest risks. The standard specifies a process that can be followed by all links in the supply chain so that information about the system can be communicated using common terminology and system parameters.

The Safety requirements for different electronic systems used in different sectors are derived from the parent IEC61508 standard requirements by refining the requirements to be specific to the application and environment of the sector's electronic systems. Several Standards derived from IEC61508 are as below:

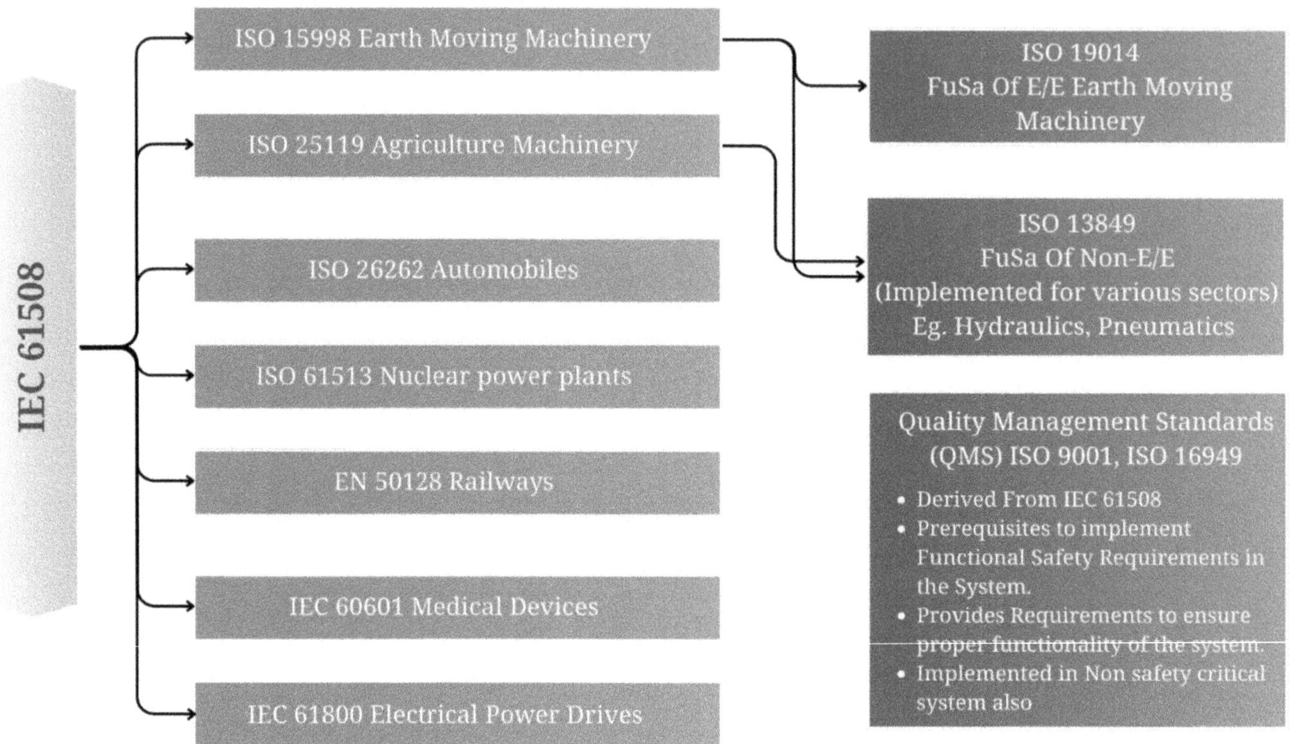

Now, let us look in detail about the Functional Safety of Earth Moving and Agricultural Machinery for a walk-through the standards other than ISO26262.

FuSa Standards of Off-highway Machinery

Since we have covered so far the ISO26262, the automotive standard that is applied for maximum vehicles that run in the Highway, lets us have a walkthrough at few of the Off-highway machinery standards, say agricultural and earth moving machinery standards, in this chapter.

The standard applicable for agricultural vehicles are ISO25119 and the standard applicable for earth moving machinery are ISO19014. The above standards cover the safety of E/E components inside the machinery of respective domains. The standard ISO13849 is applied to the Non-E/E components of the machinery of several sectors, in our case, the agricultural and earth moving machinery.

Standards	In Scope of Standard	Out of Scope of standard
ISO25119 - Safety of Agricultural Machinery	Electrical, Electronics and programmable electronic systems (E/E/PES) for agricultural machines	Hazards due to non-safety related control system and Non-E/E/PES systems such as hydraulics, pneumatics. Also the hazards caused from the system itself due to Electric shock, fire etc.
ISO19014 - Safety of Earth Moving Machinery	Hazards caused due to E/E/PES failure in control system used in earth moving machinery	Hazards due to non-safety related control system and Non-E/E/PES systems. Also the hazards caused from the system itself due to Electric shock, fire etc.
ISO13849 - Safety of Non-E/E components	Supports all kinds of technologies, mainly non-E/E/PES and wide range of machinery used for various sectors	Hazards due to non-safety related control system and Non-E/E/PES systems. Also the hazards caused from the system itself due to Electric shock, fire etc.

ISO25119 - Functional Safety of Tractors and Machinery for forestry and agriculture

The standard was published in 2010, updated in 2018, applicable to Electrical, electronic. Programmable electronic systems (PES) of agricultural machinery.

The ISO25119 standard defined Objectives based on risk. The risks are identified and eliminated using Systematic Approach The standard provides consistent document requirements to meet the safety level.

The safety level or the risk level in ISO25119 standard is estimated by AgPlr - Agricultural Performance Level. AgPlr has five levels a,b,c,d,e with 'a' being lowest risk level with lower safety requirements and with 'e' being the highest risk level with critical safety requirements.

Parts of ISO25119

ISO 25119 consists of four parts.

ISO 25119 PARTS

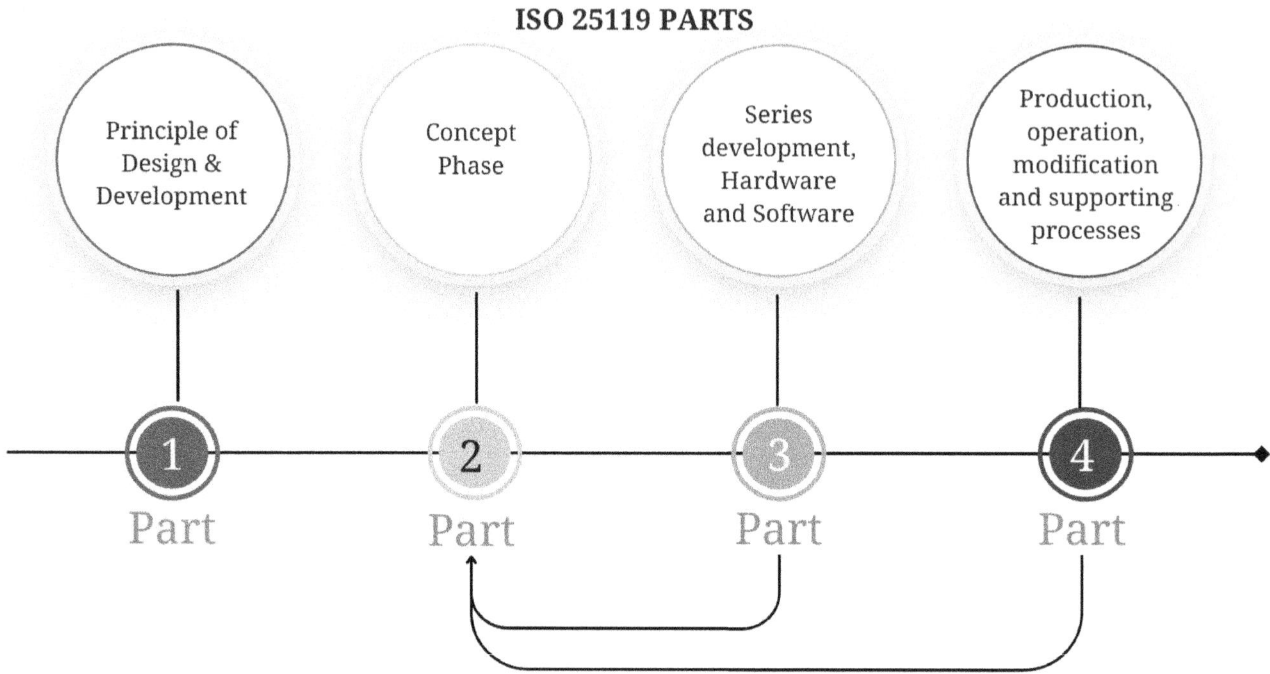

ISO 25119 is a standard related to functional safety for agricultural and forestry machinery. It specifies safety requirements and measures to be taken during the development and operation of these types of machinery. The standard comprises four parts:

- ISO 25119-1:2010 - Tractors and machinery for agriculture and forestry - Safety-related parts of control systems - Part 1: General principles for design and development: This part provides an overview of the general principles for the design and development of safety-related control systems in agricultural and forestry machinery.

- ISO 25119-2:2010 - Tractors and machinery for agriculture and forestry - Safety-related parts of control systems - Part 2: Concept phase: This part specifically deals with the concept phase of the development process for safety-related parts of control systems in agricultural and forestry machinery.

- ISO 25119-3:2010 - Tractors and machinery for agriculture and forestry - Safety-related parts of control systems - Part 3: Series development, hardware, and software: Part 3 focuses on the series development phase of safety-related control systems, covering aspects related to hardware and software.

- ISO 25119-4:2019 - Tractors and machinery for agriculture and forestry - Safety-related parts of control systems - Part 4: Production, operation, modification, and supporting processes: This part addresses the production, operation, modification, and supporting processes for safety-related control systems in agricultural and forestry machinery.

Please note that standards are periodically updated, so it's always a good idea to refer to the latest versions and editions to ensure compliance with the most recent requirements.

ISO19014 - FuSa of Earth Moving Machinery

This document provides a methodology for the determination of performance levels required for earth moving machinery (EMM) and the requirements to be implemented in the safety control system to meet the determined performance level.

Machine Control System Safety Analysis (MCSSA) is a systematic process used to determine the level of risk reduction required for hazards associated with control systems in a machine or equipment. The purpose of MCSSA is to ensure that Safety Control Systems (SCS) are designed and implemented to mitigate risks to an acceptable level.

Here's how the process generally works:

- Hazard Identification: Identify all the hazards associated with the machine or equipment, including potential scenarios where the control system may fail to perform its intended safety function.

- Risk Assessment: Evaluate the severity and likelihood of each identified hazard to determine the level of risk associated with them.

- Performance Level (PL) Determination: Assign a Performance Level (PL) to the safety functions of the control system based on the risk assessment. The PL indicates the required level of risk reduction for each safety function.

- Validation and Verification: Ensure that the control system is designed, implemented, and validated to meet the required Performance Level. This may involve using appropriate safety components, complying with safety standards, and conducting safety tests.

- Documentation: Maintain detailed documentation of the MCSSA process, including hazard analysis, risk assessment results, safety requirements, and verification records.

By conducting MCSSA, engineers and manufacturers can identify the necessary risk reduction measures and design control systems that comply with safety standards and regulations to protect users and operators from potential hazards. This approach is critical in ensuring the safe operation of machinery and equipment.

Released parts

Part 1 - Methodology to determine safety related parts of the control system and performance requirements

Part 3 - Environmental performance and test requirements for electronic and electrical components used in safety related parts of the control system

Part 4 - Design and evaluation of SW and data transmission of safety related parts of control system

Part 5 (Technical Specification) - Table of Performance levels (This part provides pre-defined MPL rating for safety control systems that are already in the field)

Parts under Development

Part 2 - Design and evaluation of HW and architecture requirements for safety related parts of the control system

Risk Estimation

Risk level in this standard is expressed by MPL (Machine Performance Levels): a,b,c,d,e

- a is the lowest risk level with less strict safety requirements

- e is the highest risk level with most strict safety requirements

- Determined by Machine control system safety analysis (MCSSA), which is similar to HARA in ISO25119

ISO13849 - Safety of Machinery

ISO13849 is a Generic Safety Standard applicable for a wide range of Machinery. This standard includes assessment of Non-electronic Components and due to this fact, this standard is applied across various sectors.

This standard is applied to Non-E/E components of machinery in sectors where there are specific standards for E/E and the defined specific standard does not cover non-E/E/PES components in the safety critical machinery of the sector.

Released parts

Part 1 - General principles of design for Safety related parts of control system

This part defines Determination of Performance level, System and SW design methods

Part 2 - Validation requirements for Safety related parts of control system

This part defines Validation by Analysis and testing methodologies

Risk Estimation

This standard mentions risk level by Performance Level (a,b,c,d,e) with 'a' being the lowest risk level and 'e' being the highest risk level. The risk level is determined by Machine control system safety analysis (MCSSA), defined by the ISO13849 for non-E/E/PES components.

Terms used for FuSa Hardware and Software

Terms used for FuSa Hardware

Safety Control System (SCS) – A system which can fail in a way that can create hazard in a safety critical system. Example: SCS for propulsion may include throttle, gear shift, start/stop control etc.

Unit of Observation (UoO)/ System of Interest – Electrical, electronic and electrically-programmable system or function and its scope, context and purpose

Safety related part of Control System (SRP CS) – Part of SCS that responds to safety related input signals and generates safety related output signals.

Machine Performance Level (MPL) / Performance Level (PL) / Agricultural Performance Level (AgPL) – Used to describe the performance level required from safety related part of control system

Machine control system safety analysis (MCSSA) / Hazard Analysis and Risk Assessment (HARA) – Risk analysis method used to determine the performance level of the machine

Common Cause Failure – Multiple failure within a system of interest resulting from a single event. Failures are not consequences of each other.

Mean Time to dangerous failure (MTTFd) – Average value of the expected time to a dangerous failure

Diagnostic Coverage – Fraction of probability of detected dangerous failures to probability of total dangerous failures

Terms used for FuSa Software

Application Software – Software specific to application, implemented by the machine manufacturer, and generally containing logic sequences, limits and expressions that control the appropriate inputs, outputs, calculations and decisions necessary to meet the SRP CS requirements.

Embedded Software – Software that is part of the system supplied by the control manufacturer which is not accessible for modification for the user of machinery such as Firmware, system software

Limited Variability Language – type of language that provides the capability of combining predefined, application-specific library functions to implement safety requirement specifications. Example: PLC

Fixed Variability Language – type of language that provides the capability of implementing a wide variety of functions and applications. Example: C, C++, Assembler

Common Functional Safety Lifecycle

We will take the development life cycle defined by ISO25119 as it defines the overall life cycle that will be followed for the off-highway machinery with maximum steps. Some of the activities may not be required by ISO13849 and ISO19014 standards.

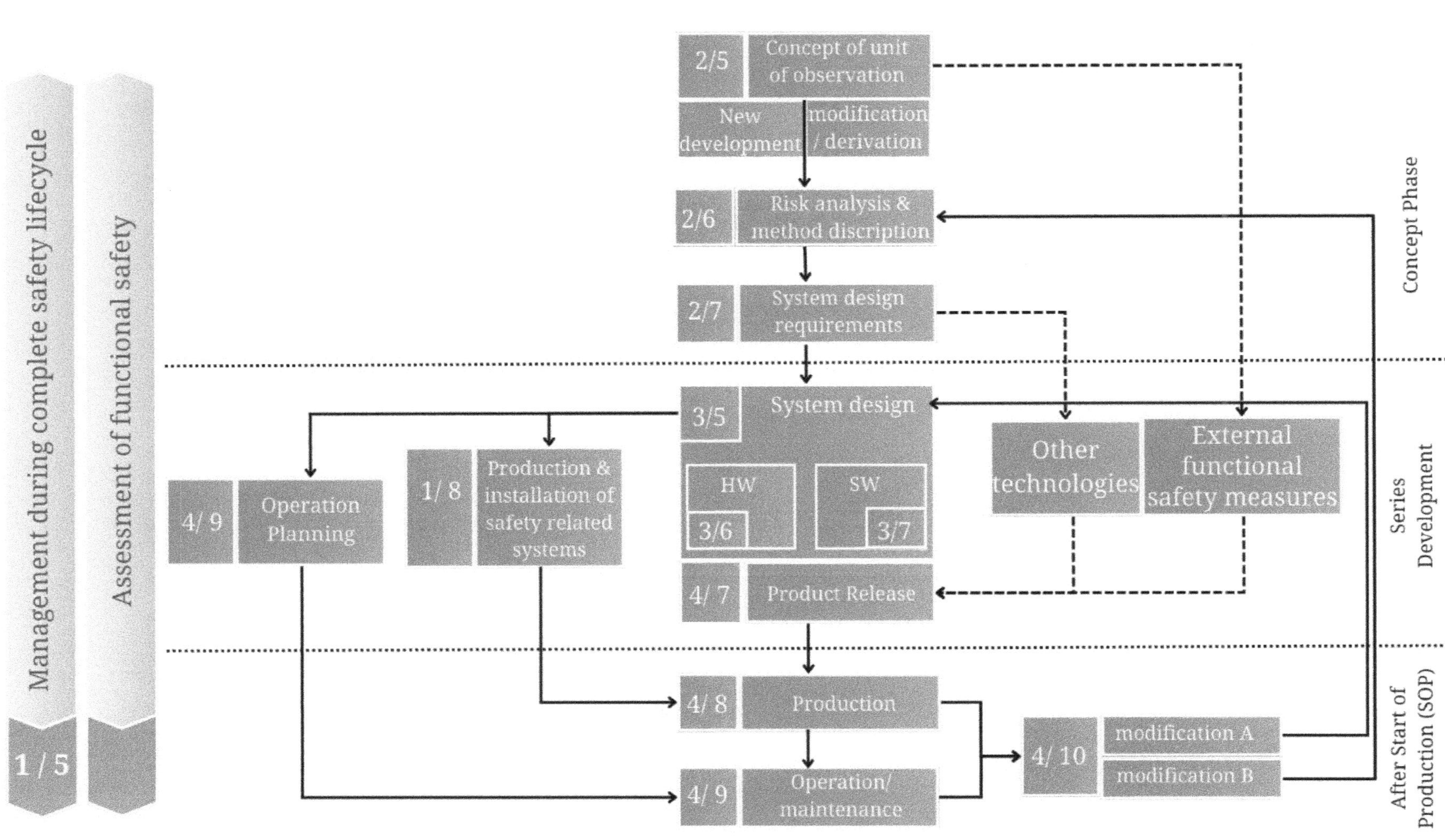

Major steps in the above lifecycle are as follows:

- Functional Safety management

- System Definition

- Risk Assessment

- Safety concepts

- Hardware and software design

- Safety verification and validation

- Safety audits and assessments

- Product Release and activities after production

We will walk through each step in detail.

Functional Safety Management

Safety Culture – cultivating a company-wide safety culture is the most important step when establishing the functional safety process within an organization. A proper safety culture necessitates the encouragement to stop a project, if, for instance, a safety violation occurs. Likewise, proper authority must be given to those responsible to uphold the safety requirements. This presumes sufficient resource quantities are allocated to projects, and they are adequately trained.

Competence Management – Competency management becomes essential to match and track the respective skill level to the allocated Roles and Responsibilities in the project. A matrix can and should be devised to identify the criteria required for the resources assigned, which may include "non-functional safety" skills, such as product domain knowledge, experience launching programs for a particular function, management expertise, and post-launch capability.

Functional Safety Plan – The safety plan is the crucial component to effectively track safety activity progression. The safety plan also consists of the details of the project hierarchy, document databases and the safety requirement traceability process. Oftentimes, the safety plan takes the form of a timeline, since it must either be referenced within or is even integrated within the project plan itself. As with a project plan, the safety plan is a living document throughout the entire project, requiring regular updates.

Supplier selection and monitoring – Supplier Monitoring has to be done via two main methods: Technical meetings to be held to discuss and clarify requirements, technical questions, problems, and quality issues. A dedicated list of open items and manage the points until closure shall be maintained. And then meetings are to be held to discuss the management part, where the progress against the schedule, check effort and costs (if it is a time and material contract) are discussed, and also with change requests, forecasts, and risks will be addressed.

Risk Assessment to Determine AgPLr, PLr or MPLr

The safety rating of a system is given by the Performance Level rating corresponding to the respective standard. This is determined by risk assessment of the function achieved by the system and the safety warning measures that will be expected from it by the user to attain a reasonably safe environment. The risk assessment is a scenario based analysis that will be performed for all the environments in which the system is expected to safely operate. This assessment is known as Hazard and Risk assessment (HARA) or Machine control system safety analysis (MCSSA)

Determination of Agriculture Performance Level rating (AgPLr) - ISO25119

1. Determines risks associated with the vehicle/machine level functions that are applicable for vehicles employed for agricultural activities.

   Example: Steering in the wrong direction of a tractor, Propelling when in neutral position, Braking when not intended.

2. Cross-functional team (Hardware, Software, Testing) provides valuable input to the HARA analysis.

3. AgPLr is a combination of Severity of harm, Probability of Occurrence of harm and Controllability.

a. Severity – S0(No injuries), S1(light to moderate injuries), S2(Severe - survival probable), S3(Severe disability)

b. Exposure – E0(Not likely), E1(Rare Event), E2(Sometimes), E3(Often), E4(Frequently)

c. Controllability – C0(Easily controllable), C1(Simply Controllable), C2(Mostly controllable), C3(Not controllable)

S	E	C	AgPlr
S0	-	C3	QM
S1	E2	C3	a
S2	E2	C3	b
S2	E3	C3	c
S2	E4	C3	d
S3	E4	C3	e

Determination of Performance Level rating (PLr) - ISO13849

1. Determines risks associated with the non electric and electronic parts that are applicable for realizing safety relevant function in the vehicles.

   Example: Under steering due to hydraulic failure

2. PLr is determined by Severity of injury, Frequency and/or exposure to hazard and possibility of avoiding the hazard or limiting harm

a. S1 (Reversible injury), S2 (Irreversible injury)

b. F1 (Short exposure time), F2 (Long exposure time)

c. P1 (limiting the harm is possible under specific circumstances), P2 (Not likely to avoid)

Severity	Exposure	Possibility of limiting hazard	PLr
S1	F1	P1	a
S1	F1	P2	b
S2	F1	P1	c
S2	F2	P1	d
S2	F2	P2	e

Determination of Machine Performance Level rating(MPLr) - ISO19014

3. The Machine Control System Safety Analysis (MCSSA) is used to determine the machine performance levels (MPLr) for Earth Moving Machinery. Eg: Articulated frame dumpers

4. MPLr is a combination of Severity of harm, Exposure and Controllability of the situation.

a. Severity – S0 (no significant injury), S1 (Minor injury), S2 (Severe injury), S3(Fatality)

b. Exposure – E0 (E<1%), E1 (1%<E<10%), E2(E>=10%)

c. Controllability – C0(High), C1(Medium), C2(Low), C3(No controllability)

5. Part5 of this standard has industry accepted predefined MPLr for machine functions of earth moving machinery

Severity	Exposure	Controllability	MPLr
S0	-	C3	QM
S1	E2	C1	a
S2	E2	C2	c
S3	E1	C1	b
S3	E2	C3	e
S3	E1	C3	d

Safety Goals

The Safety goals are Top level safety objectives of the system of interest. These are the output of the HARA or MCSSA analysis of the system. Every non-QM hazardous situation from HARA should be a safety goal.

Examples:

- Avoid loss of intended function

- Avoid uncommanded activation of intended function (steering/Brake)

Safety Concepts

Functional Safety Concept

The Functional Safety Concept is the workproduct that defines the Functional Safety Requirements (FSRs) that are derived from the Safety Goals defined for the system. The FSRs are requirements that shall be implemented at each system block level to satisfy the Safety goal. An FSR shall be associated with following attributes:

- AgPlr - a,b,c,d,e

- Diagnostic Coverage - Low, Medium, High

- Mean Time to Dangerous Failure - Low, Medium, High

- Software Required Level - B,1,2,3,4

- Hardware Category required - B,1,2,3,4

FSR Examples:

- The presence of an operator shall be detected

- Override features shall be used to disable the operational feature in case of an error

Technical Safety Concept

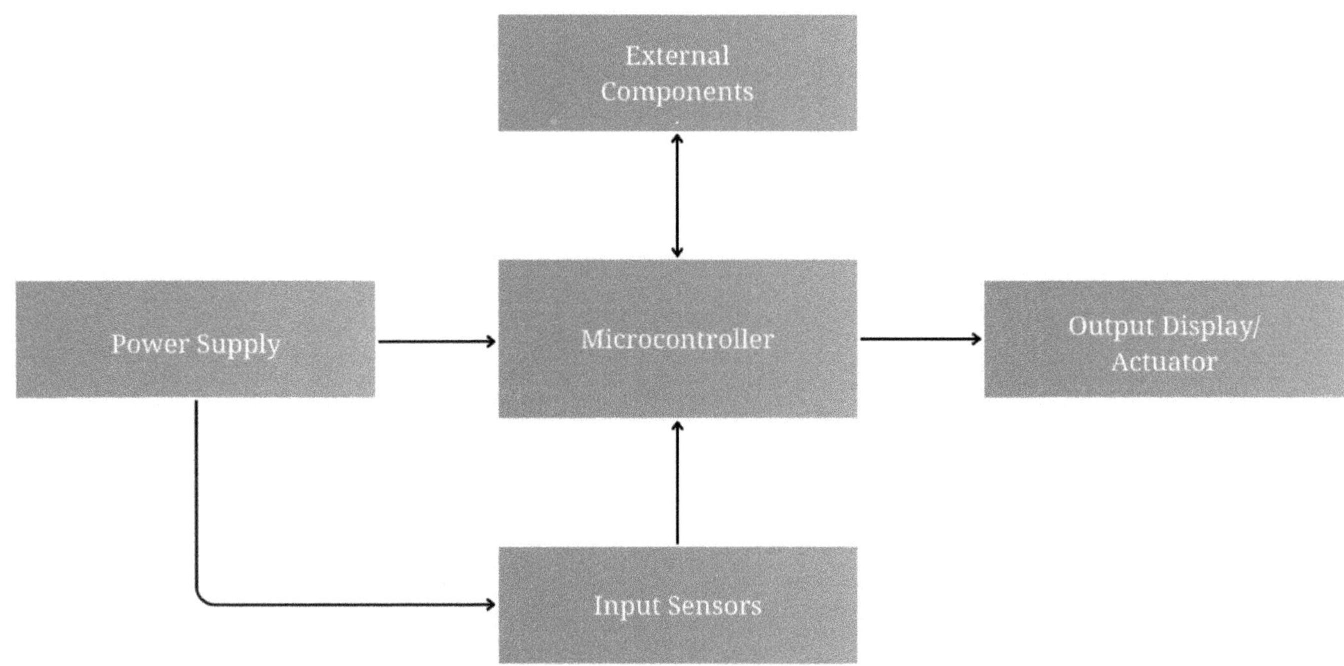

The Technical Safety Concept is the document that includes Technical Safety Requirements (TSRs), that are derived from FSRs. The TSRs are safety requirements that satisfy the FSRs at Hardware block level, as per the above figure.

This document includes Hardware Architecture with safety requirements for individual hardware components, interfaces between the components. It also contains safety requirements for the software that has to be implemented to prevent, mitigate and reach the safe state of the system.

The Hardware and Software safety requirements are derived from the TSRs. Safety Analysis methods such as fault tree analysis, Failure modes and effect analysis shall be used to identify faults and their mitigation actions.

Example TSR: The Microcontroller shall implement undervoltage and overvoltage monitoring for the power supply.

Safety requirements at various project phases

The Safety requirements that are defined by Technical Safety Concept will be refined to separate Hardware and Software requirements that will be implemented during the design of respective elements. Hence, once the TSRs are provided by the Safety engineer, the requirement shall be handled by the Hardware and Software engineers by the upcoming activities.

Hardware Design Key Activities

Each TSRs will be defined with Required Hardware category, MTTFd, and Diagnostic Coverage. This shall be satisfied by the Hardware requirements. The Hardware team is responsible for the below activities.

- Selection of appropriate hardware category of an element(Cat B, Cat 1, Cat 2, Cat 3)

- Calculation and verification of Diagnostic Coverage

- Calculation and verification of MTTFd

- Component MTTFd from standards and databases such as MIL_HDBK-217F, SN29500

- Analysis of Common cause failures (if required)

Hardware Categories

The various hardware categories can be detailed as below:

- No monitoring required of category B/1. For Cat B/1, we may use well-tried PIU components, if needed.

- For Cat 2, it has some logic detecting mechanisms that could detect and inform the user on violation of SG. TE- testing element

- For Cat 3/4, it will have redundancy to eliminate Single point faults by cross-monitoring of the logic

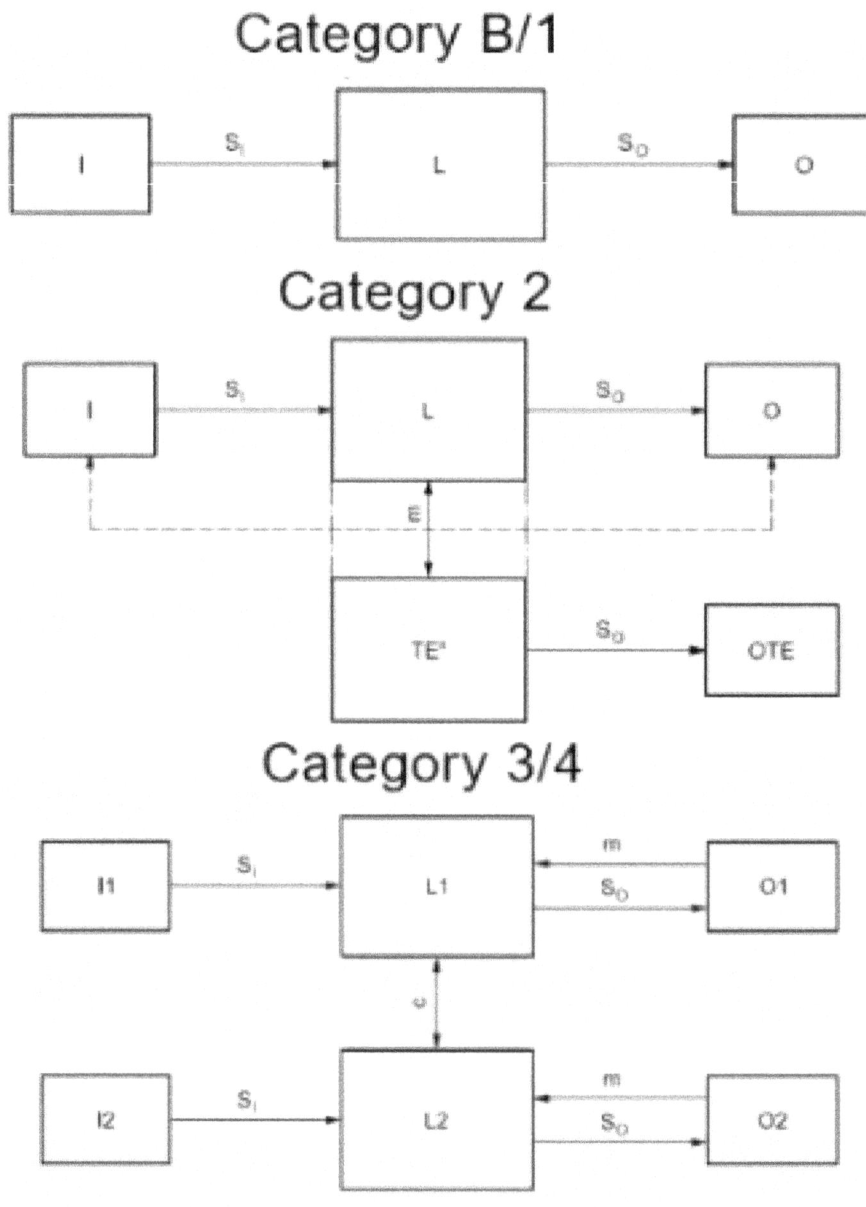

MTTFd and Diagnostic Coverage

The Mean Time To Failure and the Diagnostic coverage is provided by the manufacturer or supplier of hardware components. Incase it is not available, it is calculated by using a reliability standard, for example, SIEMENS, IEC62380, MIL-HDBK 217

Common Cause Failure (CCF)

Common Cause failure analysis is required for Category 3 and 4. Common cause failures are defined as a single failure event where multiple failures are caused within a system of interest resulting from the single event, where these failures are not consequence of each other. Example: Over voltage, over temperature, Ground Noise.

The common cause failures can be eliminated by taking suitable measures. Some of the measures include separation of signal and power path, diversity in implementation of the main path and the redundant path, design robustness increase against environmental stress.

Software Development Key Activities

The below is the software safety lifecycle model defined by IEC61508. The standards derived from IEC61508 also follow a similar development cycle.

This is a V-model where the left side of the V are development activities where you refine from higher requirements and track the same to be implemented in the code. Throughout the right side of the model the written code will be tested to satisfy the defined requirements

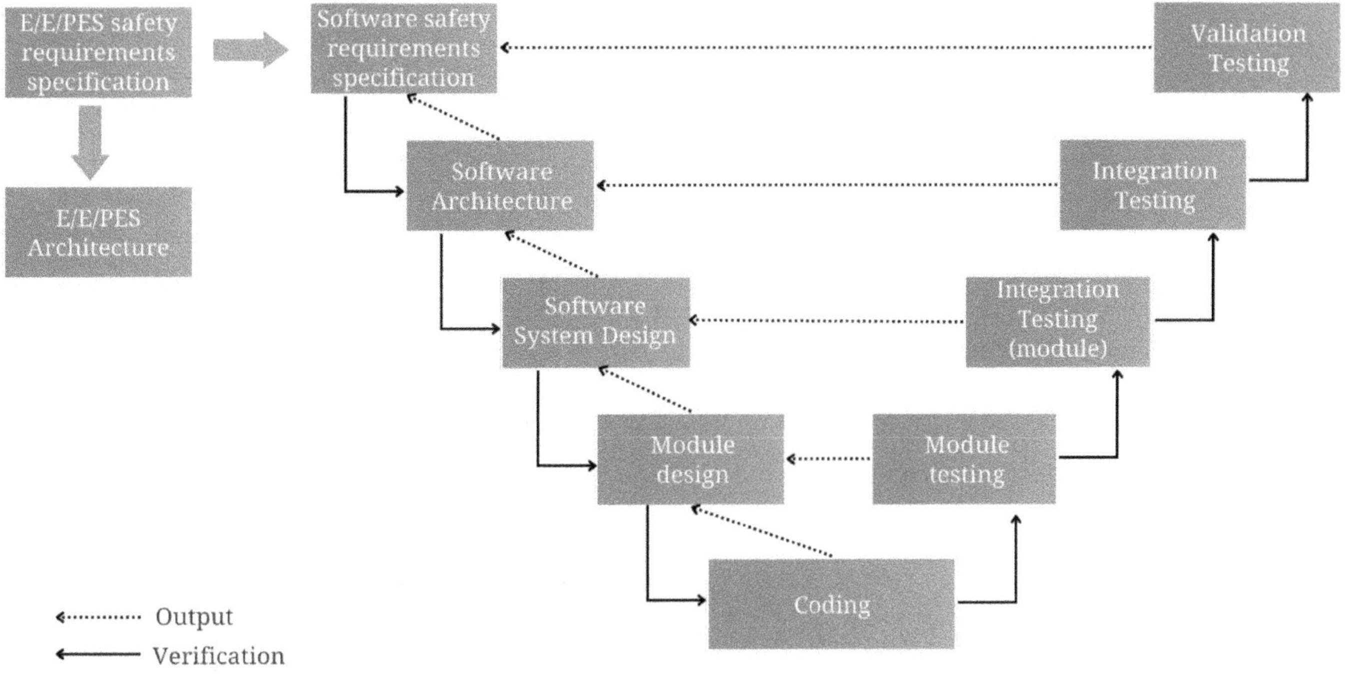

For ISO25119 and ISO19014, below are the major key software activities:

- Software Safety Requirement and Analysis (validated by Software Safety Testing)

- Software Architecture and Design (validated by Integration Testing)

- Software Component Design and Implementation (validated by Software Component Testing)

Note: () represents corresponding test activities

Example of Software Safety Concepts

Before diving into the details of Software Key activities, let us discuss a couple of examples of Software Safety Concepts to understand the need and purpose of safety requirements implemented in Software.

Example 1

Let us take the below example of a microcontroller where the programming memory is partitioned to A and B and two different functions are programmed independently in the partitioned memories.

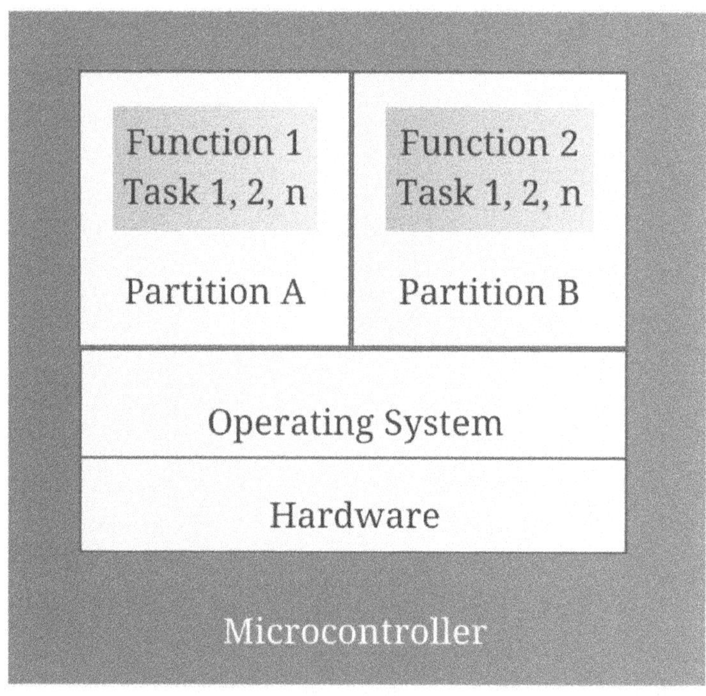

In this type of implementation, we may incur failure effects like memory corruption, blocking of partitions, wrong processor execution time, wrong communication peer, corruption of I/O interface, etc.

Safety mechanisms can be integrated into the software to counteract the above mentioned failures. Some of the safety mechanisms that can be implemented are:

1. For verification of communication

    a. Unambiguous bidirectional communication object

    b. Strictly two unidirectional communication objects

    c. Asynchronous data communication

2. For allocation of processor execution time

    a. No priority based scheduling

    b. Time slicing method

3. For Allocation of System Resources

    a. Memory protection mechanisms

    b. Verification of safety critical data

    c. Static analysis

    d. Static allocation

Example2

Let us take another example as below:

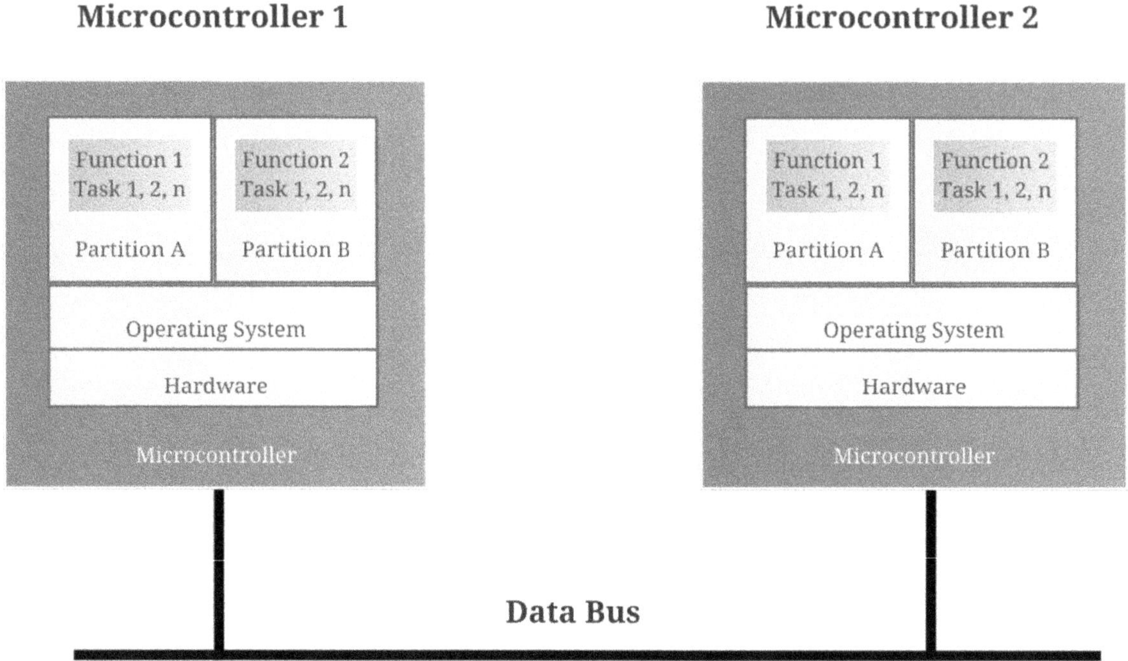

Two microcontrollers connected via a data bus as above can be implemented safely without failure due to communication such as corruption, signal distortion, collision etc. by implementing respective software safety mechanisms.

Few of the mechanisms that can be implemented in the above design for realizing a safe system are:

1. Verification of communication

    a. Alive counter, CRC, sequence number

    b. Message repetition, Watchdog

2. Bus Allocation

    a. Bus guardian, Mini-slotting

    b. Time triggered data bus

Software Safety Development Planning

The Software Safety development plan determines individual phases of software development such as

- Software Safety Requirements

- Software Architecture

- Software Design

- Software Coding

- Software integration and testing

- Software Validation Test Plan

Software Safety Requirements

Software Safety Requirements (SSRs) are derived from Technical Safety Requirements. These requirements shall inherit the performance levels (PLr) and software required levels(SRL). The SSRs are verified if they are consistent with the TSRs and shall be confirmed with the safety traceability report.

The SSR typically includes:

- Software Architecture

- Attributes of system hardware

- Response time of Safety-related functions

- External Interfaces such as communication

- Physical requirements and environmental conditions as far as they affect the software

- Requirements for safe software modification

Typical examples for SSRs:

- At power up, read sensor output to detect voltage transitions

- Monitor ADC inputs to microcontroller to detect any error value

Software Architecture Design

The software architecture of a system represents the design decisions related to overall system structure and behavior. The Software architecture requirements are expected to satisfy the PLr by refining the defined SSRs.

The key methods to define an efficient software architecture are by using:

- Computer aided specification tools

- Cyclic behaviour and time triggered architecture

- Review recommendations (walkthrough and inspection)

- Ensuring traceability between System and Software Safety Requirements

Software Component Design and Code

Software component design and code development requirement specifies in detail the behaviour of the safety-related software components that are prescribed by software architecture. The aim is to generate a readable, testable and maintainable source (code, model, etc.,) suitable to be translated into object code. The development of software code shall follow design and coding guidelines.

The software component compliance with the guidelines can be confirmed by following methods:

- Use of suitable tools and programming languages

- Informal, semi-formal, formal, defensive, structured programming

- Limited use of interrupts and defined use of pointers

- Inspection and walkthrough of software design

Software Integration and Testing

Software Integration and Testing are performed when integrating individual software components into a complete Embedded Software.

The Software Integration strategy includes:

- The steps to be taken for integrating the individual SW components hierarchically;

- Functional dependencies that are relevant to the Software integration

Key methods by which the software integration testing is performed:

- Functional or black box testing

- Equivalence class and input partition testing

- Performance testing

Software Safety Testing

The objective of Software safety testing is to confirm that the software requirements are correctly realized by the embedded software. It is part of Safety validation.

The main test environments to perform Software Safety Testing are:

- Hardware-In-Loop tests

- Tests within the ECU networks

- Vehicle testing

Validation and verification of safety requirements

Verification is the process of comparing a safety circuit design and its components to the appropriate safety standards to determine if they meet or exceed the safety level determined by the risk assessment. For example, after the safety circuit is designed and before it is installed, SISTEMA, a widely-known software tool, can be used to accomplish a safety system verification for the control standard of EN ISO 13849-1. The software allows the engineers to model the structure of the safety devices in the system and input the relevant parameters to determine the achieved Performance Level (PL) of the safety system.

Validation is the process of checking the possible fault conditions of the safety system after it is installed to make sure it functions to the safety level required by the risk assessment. This process may involve lifting wires, shorting contacts, removing power, etc., to simulate the possible fault conditions.

Assessment of Functional Safety

The assessment of Functional Safety includes judgements made by assessors to ensure that functional safety is achieved by System of interest via reviews, safety audits and safety assessments. It ensures that all the activities and documentation for the Safety lifecycle have been completed as per requirements. The organizational unit responsible for functional safety of the machine carries out the assessment of Functional Safety. The planning for assessment of Safety activities should be a part of the project Safety plan and shall be managed by the Functional Safety Manager.

The main work products that will be analyzed will be

- Hazard analysis and Risk Assessment

- Functional Safety Plan

- Verification and Validation of safety requirements

- Functional and Technical Safety Concepts

- Safety Analysis techniques and test cases

- Safety Case

Case Study - Automated Steering System of a Tractor

Introduction

The Automated Steering System (ASS) is an innovative technology integrated into tractors to enhance their efficiency, precision, and safety. This case study focuses on the implementation and benefits of an ASS in a tractor, detailing its functionalities and the advantages it offers to farmers and agricultural operations. For a better understanding of the importance of functional safety, let us discuss the use case of the automated steering system. Many users in dry-field farming have backed the tractor with an autonomous steering system

Problem Statement

Traditional manual steering in tractors can be physically demanding and may lead to inaccuracies and overlaps during field operations. This case study aims to address these issues and explore the benefits of adopting an Automated Steering System.

Objectives

To improve the precision and accuracy of tractor steering during various field operations.

To reduce operator fatigue and enhance overall operational efficiency.

To optimize fuel consumption by minimizing overlapping in field tasks.

To increase the safety and reliability of tractor operations.

To evaluate the return on investment and cost-effectiveness of implementing an Automated Steering System.

Implementation and System Description

The Automated Steering System in the tractor comprises various components, including Global Navigation Satellite System (GNSS) receivers, an electronic control unit (ECU), hydraulic steering control, and an operator interface. The GNSS receivers receive signals from satellites to determine the tractor's precise position and heading. The ECU processes this data and provides real-time steering commands to the hydraulic system, controlling the tractor's steering angle.

Functionalities

Auto-Guidance: The ASS allows the tractor to follow predefined paths with sub-inch accuracy, minimizing overlaps and optimizing field coverage. It supports straight-line guidance, curved paths, and contour following.

Headland Management: During turning maneuvers at the field edges, the ASS automates the steering to maintain consistent turning radius and reduce operator workload.

Line Acquisition: The system can automatically engage the guidance once the tractor reaches the desired starting point, enabling easy integration with different field shapes.

Correction Signals: Real-time differential correction signals from satellite-based correction services improve the positioning accuracy, resulting in precise tractor path following.

Benefits

Increased Efficiency: The ASS reduces manual errors and overlapping, leading to more efficient field operations and optimal use of resources.

Operator Comfort: With automated steering, the operator experiences less physical strain and can focus on other tasks like monitoring implements or making necessary adjustments.

Time Savings: The system enables continuous operation without the need for manual course corrections, reducing turnaround times at field boundaries.

Improved Safety: Precise steering reduces the risk of accidents, especially during night or low-visibility conditions.

Economic Impact

The initial investment cost for an Automated Steering System is offset by the gains in operational efficiency, reduced fuel consumption, and improved crop yield due to precise operations. A cost-benefit analysis reveals the return on investment and validates the economic feasibility of the system.

Conclusion

The implementation of an Automated Steering System in a tractor offers significant advantages in terms of precision, efficiency, and safety. By reducing operator fatigue and optimizing field operations, the technology contributes to increased productivity and improved yield in agricultural practices. The case study emphasizes the importance of integrating cutting-edge technologies into agricultural machinery to meet the evolving demands of modern farming practices. Therefore, based on ISO25119, a new safety evaluation was developed and used to evaluate the autonomous steering system. Malfunction or hazards analysis needs to be conducted compliant to the Functional Safety standard to determine the consequences of an electronic failure. Based on this analysis, the agricultural performance level of the function with corresponding Safety goals will be defined. The system shall be developed correspondingly to meet the AgPlr and shall satisfy all the process and technical requirements as per ISO25119 to eliminate all the risks of this automated system.

GNSS receiver
Trimble
Display
Autopilot controller with
integrated IMU gyroscopes
CASE iH
MX305
Front wheel angle
turn sensor
Hydraulic valve with hose kit

References

1. Presentation on Driver assistance systems (DAS) Transmitted by the Government of Israel By Economic Commission for Europe • October 2014

2. Perception-Response Time to Unexpected Roadway Hazards By Paul L. Olson and Michael Sivak; Human Factors: The Journal of the Human Factors and Ergonomics Society • February 1986

3. ISO 26262-12:2018 Road vehicles — Functional safety By International Organization for Standardization

4. Designing ISO 26262-Compliant HMI Graphics and Storage for Automotive Clusters By Pritesh Mandaliya • October 2019

5. Electrification of automobile by our power management ICs By Nisshinbo Micro Devices Inc.

6. Quality management in Bosch - Fault tree analysis FTA By Robert Bosch GmbH • August 2015

7. Understanding the implications of the ISO 26262standard on electronic systems development By Optima • October 2019"

8. New Rear Occupant Alert Reducing Child Heat Hazards By Hyundai Motor • October 2017

9. An introduction to ASIL decomposition and SIL synthesis By Texas Instruments • April 2019

10. The Hardware/Software Interface: Where We've Been, and Where We're Going By Anupam Bakshi • October 2018

11. Let's Talk About Configuration Management and ISO 26262 By Jonathan Moore • August 2019

12. Qualifying Software Tools According to ISO 26262 By Mathworks & TUV SUD

13. Why is Software Tool Qualification Indispensable in ISO 26262 Based Software Development? By Embitel • November 2020

14. Clearing the Fog of ISO 26262 Tool Qualification By Joseph Dailey and Jake Wiltgen • April 2022

15. CASE STUDY - Honda 1800 Goldwing Runaway By MSG Group

16. ISO 26262 vs. SOTIF (ISO/PAS 21448): What's the Difference? By Hanna Taller • June 2022

17. What is SOTIF? (Safety of the Intended Functionality) ISO/PAS 21448 By Adam Saenz • March 2022

18. The emerging role of cybersecurity in the automotive sector By Manav Kapur in Voices, TOI • September 2022

19. 10 Automotive Cybersecurity Examples to Know By BuiltIn • July 2022

20. ISO 25119:2018 Tractors and machinery for agriculture and forestry — Safety-related parts of control systems By International Organization for Standardization

21. ISO 13849:2015 Safety of machinery — Safety-related parts of control systems By International Organization for Standardization

22. ISO 19014:2018 Earth-moving machinery — Functional safety By International Organization for Standardization

23. Updated ISO 25119 Series to Improve Farm Safety By JANE BOLER • November 2018

24. ISO 25119: Software Development for Tractors and Machinery for agriculture and forestry By HEICON Global Engineering

25. Safety Functional Requirements for Robot Fleets for Highly Effective Agriculture and Forestry Management By Conference: International Conference of Agriculture Engineering (CIGR 2012)At: Valencia, Spain • July 2012

26. Implementing Functional Safety to off-highway equipment - Agriculture and Forestry machines By TATA Elxsi

"My Journey To Remember, To Acknowledge...
My Journey That Thought me And Defined me."